Criticizing Science

GLOBAL STUDIES IN MEDICINE, SCIENCE, RACE, AND COLONIALISM

Ahmed Ragab, *Series Editor*

Criticizing Science

STEPHEN JAY GOULD AND THE STRUGGLE FOR AMERICAN DEMOCRACY

MYRNA PEREZ

Johns Hopkins University Press
Baltimore

Printed in the United States of America on acid-free paper

2 4 6 8 9 7 5 3 1

Johns Hopkins University Press
2715 North Charles Street
Baltimore, Maryland 21218
www.press.jhu.edu

Library of Congress Cataloging-in-Publication Data

Names: Sheldon, Myrna Perez, author.
Title: Criticizing science : Stephen Jay Gould and the struggle for American democracy / Myrna Perez.
Description: Baltimore : Johns Hopkins University Press, [2024] | Series: Global studies in medicine, science, race, and colonialism | Includes bibliographical references and index.
Identifiers: LCCN 2024010353 | ISBN 9781421450155 (hardcover) | ISBN 9781421450179 (ebook)
Subjects: LCSH: Gould, Stephen Jay. | Science—United States—Public opinion. | Right and left (Political science)— United States.
Classification: LCC Q172.5.P82 S54 2024 | DDC 352.7/45—dc23/eng/20240425
LC record available at https://lccn.loc.gov/2024010353

Special discounts are available for bulk purchases of this book. For more information, please contact Special Sales at specialsales@jh.edu.

For Garnet, Indigo, and Seth

CONTENTS

Acknowledgments *ix*

INTRODUCTION 1

1 Racism and the Promise of a Democratic Science 22

2 Sociobiology and Gender Trouble 50

3 Evolutionarily Engineered Sexism 81

4 Scientists Confront Creationism 107

5 The Accidental Creationists 133

6 Is a Democratic Science the End of Western Secularism? 159

EPILOGUE Why We Can't Wait for a Democratic Science 185

Notes *193*

Index *237*

ACKNOWLEDGMENTS

This book stems from some of my earliest intellectual questions, while also capturing many of those I had in the ensuing two decades. I have had many fortunate moments when people and institutions welcomed me into scholarly conversations on subjects in which I had neither expertise nor training. In the omnivory of my academic pursuits I have perhaps learned from Stephen Jay Gould, who has been my constant companion over the last decade.

I am deeply grateful to Audra Wolfe, the developmental editor for this book. Her work helped me to transform the project at a crucial juncture and to find my voice in the process. I am grateful also to Ahmed Ragab, the series editor for Global Studies in Medicine, Science, Race, and Colonialism, and to Matthew McAdam, at Johns Hopkins University Press, for their enthusiasm, insight, and belief in this book.

My research and writing for this book were generously supported by funding from many institutions and organizations, including the Cutler Scholars Program at Ohio University, the Charles Warren Center for Studies in American History, the Dibner Fellowship at the Huntington Library, the Erwin Hiebert Grant in the History of Science, the Mustard Seed Foundation, the Honors Tutorial College at Ohio University, the Humanities Research Fund at Ohio University, and the Templeton World Charity Foundation.

Many archives and libraries aided me generously. I extend particular thanks to Jenny Johnson, the head archivist for the Stephen Jay Gould Papers at the Department of Special Collections, Stanford University Libraries. Jenny was an invaluable conversation partner in the early days of researching in the collection, and I remain grateful for her interest in the work. I am thankful also to the American Philosophical Society, the Huntington Library, the Library Company of Philadelphia, the Schlesinger Library on the History of Women, and the University Archives at Indiana University for support to conduct research in their collections.

I was fortunate to speak with many people whose lives intersected with this history, including Jerry Adler, Gar Allen, James Gorman, Linda Harrar, Everett Mendlesohn, Richard Lewontin, Rosamund Purcell, Frank Schaeffer, and Herbert Vetter. These conversations helped to shape and deepen my understanding of the history under my care.

I remain grateful for the early inspiration for this history that came during my time at Westmont College, as a student of Jeffrey Schloss, Richard Pointer, and Marianne Robins.

This book had its first iteration as a dissertation in the Department of the History of Science at Harvard University. Janet Browne, my doctoral adviser, was a constant source of intellectual inspiration and warm support. My doctoral committee, Andrew Jewett, Sarah Richardson, and Vassiliki Betty Smocovitis, encouraged me to strengthen my historical and theoretical analysis. A fellowship during these years at the Darwin Correspondence Project helped me to solidify my own study of Charles Darwin, and another at the Harvard Divinity School deepened my interest in science and religion. My thinking was supported by conversations with many at these institutions, especially Rosemary Clarkson, John Durant, Samantha Evans, Megan Formato, Anne Harrington, Shelley Innes, Eli Nelson, Naomi Oreskes, Alison Pearn, Lukas Riepel, Anne Secord, James Secord, Elizabeth Smith, Sally Stafford, Jenna Tonn, Lewis West, and Rebecca Woods.

The Center for the Study of Women, Gender, and Sexuality at Rice University supported this work during a two-year postdoctoral fellowship that transformed the possibilities of scholarship for me. I am especially thankful for the mentorship of Rosemary Hennessy and Fay Yarbrough and for Carly Thomsen's profound insight into the world of feminist and queer theory. I also thank the center's core faculty, including Krista Comer, Cymene Howe, Susan Lurie, and Brian Riedel.

I rewrote this book extensively after becoming a faculty member at Ohio University, where I have a joint appointment in Classics and Religious Studies and Women's, Gender, and Sexuality Studies. Both have been spaces of generosity, camaraderie, and intellectual exploration. I am grateful to each of my colleagues in these units, past and present, including Cynthia Anderson, Jim Andrews, Neil Bernstein, Tom Carpenter, Brian Collins, Cory Crawford, Fred Drogula, Steve Hays, Lynne Lancaster, Kim Little, Loren Lybarger, Eve Ng, Ruth Palmer, Nicole

Reynolds, Patty Stokes, Julie White, and Risa Whitson. Thanks also to Kristin Distel and Yashvita Kanuganti for proofreading and checking references for the final manuscript.

Many other institutions have offered me space and time to think and hold conversations in and around the work of this book. I thank the Arizona State University History of Biology Seminar for including me in two summer sessions at the Marine Biological Laboratory at Woods Hole, which aided my thinking enormously. I am grateful to James Collins, Jane Maienschein, and Karl Matlin for hosting my work at these sessions and to John Beatty and Alison McConwell for their reading and commentary on my writing in progress. At these sessions and elsewhere, I have benefited from conversations with Andrew Aghapour, Mark Borello, Henry Cowles, Chris Haufe, Patrick Forber, Michael Gordin, Kathryn Lofton, Patrick McCray, Joanna Radin, Adam Shapiro, and Tisa Wenger.

David Seposki provided incisive critique at many stages and demonstrated wholehearted faith in the project. Terence Keel offered me political solidarity and theoretical inspiration. And Erika Milam read many, many drafts with both historical expertise and steadfast encouragement.

My family has been a source of cheer, encouragement, and good will as I have written and rewritten this book at various kitchen tables, couches, and window seats over the years. My parents-in-law, Cheryl and Clark, have been unfailing in their warm interest in my efforts. Bryan Adkins and Hana Carney always gave me a corner to write in, the best snacks, and resounding legal debates. I am thankful to Derek Jones and Wolff for their warmth and spiritual steadiness. Lisa Crystal and Bryan LaPlant supplied me with cheer when I needed it most, and Lisa read and thought and wrote with me over many long years of friendship and conversation. Laura and Zack Turner hosted me for every trip I made to view the Gould Papers, and Lauren and Brian Keaney did the same during a fellowship at the Huntington Library. Betsie and Keith Cialino, Kjessie and Andre Essue, as well as Rachel Prandini and T. J. McKillop, welcomed my history talk during many weekends of fun. Betsie, Laura, Lauren, and Rachel were, and remain, my moral compass. To them I owe any possibility that the writing can be my good work in the world. And I have unending gratitude for being able to do this work with and alongside Steph Dick. I dream of the things we will make together.

Most of all I am thankful to the family with whom I share daily life. My brother, Bren, is my companion through a life of art. My mother, Terrie, makes life possible and joyful. My children, Indigo and Garnet, are the brightest of lights. And there could be no better partner in this work, or in all things, than Seth.

Portions of chapters 1 and 2 appeared in "Evolutionary Activism: Stephen Jay Gould, the New Left and Sociobiology," *Endeavour* 37, no. 2 (2013): 104–11.

Portions of chapter 3 appeared in "Claiming Darwin: Stephen Jay Gould in Contests over Evolutionary Orthodoxy and Public Perception, 1977–2002," *Studies in History and Philosophy of Science Part C: Studies in History and Philosophy of Biological and Biomedical Sciences* 45 (March 2014): 139–47, and "Wild at Heart: How Sociobiology and Evolutionary Psychology Helped Influence the Construction of Heterosexual Masculinity in American Evangelicalism," *Signs: Journal of Women in Culture and Society* 42, no. 4 (June 2017): 977–98.

Portions of chapter 4 appeared in "Science, Medicine, and Technology," in *The Routledge History of the Twentieth-Century United States*, ed. Jerald E. Podair and Darren Dochuk (New York: Routledge, 2018), 325–36, and "A Feminist Theology of Abortion," in *Critical Approaches to Science and Religion*, ed. Myrna Perez Sheldon, Ahmed Ragab, and Terence Keel (New York: Columbia University Press, 2023).

A portion of chapter 6 appeared in "The Study of Science and Religion as Social Justice," *Historical Studies in the Natural Sciences* 46, no. 4 (2016): 538–47.

Criticizing Science

Introduction

Samuel George Morton's skulls are the origin story of American science. For decades, it was possible to view the collection of more than one thousand human skulls that resides in the University of Pennsylvania Museum of Archaeology and Anthropology in Philadelphia.[1] Morton gathered them for several decades before his death in 1851, purchasing them from traders who scoured the battlefields of Europe, dug up Native American burial grounds, and robbed the graveyards of Black Philadelphians to gather evidence for his race science.[2]

Morton was a man of meticulous empirical skill. He was also a community builder. Although he seldom traveled from his home, he helped forge the American School of Ethnology, which was the young republic's first internationally recognized scientific school, through correspondence and subscriptions to his great volumes. Throwing off the strictures of the Christian sacred scripture, Morton and such men as Louis Agassiz, Josiah Nott, and George Gliddon developed a racial theory based on archaeological evidence and natural history. In their commitment to empiricism and natural knowledge these men argued that God had created the five human races as separate species. This polygenist view of human history became part of a new scientific (rather than strictly religious) justification for slavery in the United States before the Civil War.[3] Morton's particular contribution to this work was to carefully measure the skulls in his collection, using differences in skull size to bolster a claim that Caucasians were the most intelligent of the five races.

These were the racist beginnings of a professionalizing science in the United States. Scholars and activists have worked in the last several decades to confront this inheritance. Historians have uncovered the networks of plunder that enabled Morton's collection and traced the circulation of lithographic representations of the skulls through international scientific circles.[4] Most notably, in 2021 the Penn Museum announced its intention to repatriate portions of the Morton

Cranial Collection in a process modeled on the Native American Graves Protection and Repatriation Act of 1990.[5] Reckoning with the legacy of racism from Morton's scientific circle is vital and ongoing work.

Why, then, did a team of anthropologists led by Jason Lewis decide to recreate Morton's skull measurements and publish a vindication of his results in a study for *PLOS Biology* in 2011? The answer to this puzzle lies in the article's title, "The Mismeasure of Science: Stephen Jay Gould versus Samuel George Morton on Skulls and Bias."[6] The stated purpose of this study was to question Morton's status "as a canonical example of scientific misconduct," a claim established by Stephen Jay Gould in his 1981 book *The Mismeasure of Man.*[7] Gould was an evolutionary theorist who had cultivated a leftist ethos during his time as a student activist during the 1960s civil rights and anti-Vietnam movements and then brought these sensibilities with him to his professorship at Harvard in the 1970s. Gould wrote of Morton as one of many who exemplified how racial prejudice had influenced the historical development of the science of intelligence. In his discussion of Morton, Gould argued that it was not "fraud or conscious manipulation" that had led the older scientist to his claim that "blacks and Indians" were "separate species, inferior to whites," but rather the sheer power of unconscious racism in the practice of Morton's science.[8] Gould's arguments about Morton were born out of a leftist political understanding of the role of racism in American science.

What happened in the thirty years between the publication of *The Mismeasure of Man* and the publication of "The Mismeasure of Science"? Why did the *PLOS Biology* authors feel the need to defend science from accusations of racism? Telling that story is one of the purposes of this book. It begins in the early days of Gould's scientific career and in the broader context of the radical science movement and feminist science studies in the 1970s. As part of their participation in New Left political movements, including the civil rights and women's movements, these activist and academic groups claimed that demystifying science would create a better world. And in their effort to create a more democratic academy and a more just American society, they developed a critique of scientific objectivity. Alongside writing produced by such figures in early feminist science collectives as Ruth Hubbard and Marian Lowe, Gould's public science writing became a model for how to use history to critique Western science for its racism and sexism.

But alongside these conflicts, another group also targeted professional scientific expertise in their efforts to remake America. The creationists who were part of the rise of the Christian Right in the 1970s blamed Darwinian science for the downfall of biblical values in the West. Their criticism of science was part of a massive political and theological effort on the part of evangelicals to defeat secularism and return gender roles, sexual ethics, and public education in the United States to a Christian foundation. Although the 2011 *PLOS Biology* article never speaks of creationism, it is the political background that haunts the article's liberal defense of the objectivity and authority of science.

The understanding in the United States of public criticisms of science has become muddled because the instinct appears to come from both sides of the political spectrum. Although many accounts treat either right- or left-wing engagements with science, rarely, if ever, have these been considered together.[9] This history weaves them together, demonstrating how these conflicts fed into and shaped each other. Through the window of Gould's public career and private papers we see that one field—evolutionary science—simultaneously divided left-wing and liberal thinkers over race, sex, gender, and sexuality and drove the Right's religiously inflected criticisms of secularism. Technical questions about how evolution worked—its mechanisms, history, and contemporary effects—were at the heart of feminist critiques of new biological approaches to human sociality, as well as of creationist attacks on professional evolutionary biology. Ultimately, a history of evolutionary science in this period provides a tangible thread with which to follow the emergence of left-wing, liberal, and right-wing approaches to science, even as these perspectives went on to shape a range of other public controversies.

But a faithful accounting of history is not enough. Deciding how democratic science ought to be has urgent ramifications in a world in which we must decide how to vaccinate, respond to climate change, address maternal mortality, and regulate the spread of global infectious disease. It is pressing as we decide how science is used as evidence in courtrooms and whether tools developed by science can be trusted to surveil our societies. Ought we defer to the authority of professional science, as liberal science advocates argue? Or is there room for populist understandings of the nature and reliability of science? Contemplating these ethical questions is the second purpose of the book, and it is the reason this history is centered on Gould even as it continually contextualizes his career in broader scientific and political landscapes. Gould's understanding of the nature of scientific racism continues to shape our own political dialogue on science and social justice. Entering a dialogue with Gould's efforts in *The Mismeasure of Man* can help us to see both the advantages and the limitations of his political perspective on the nature of science and its relationship to democracy.

When I first read the *PLOS Biology* article on Morton's skulls, I was only a few years into my own study of Gould, lost in the papers of a thinker so multifarious that he tended to evade my youthful attempts to create a cohesive narrative about his work. But I felt I should be able to say something in response to Lewis and his team, especially since I had decided to immerse myself in a study of Gould's archive. I returned to the boxes and dutifully pored over the notes that Gould had used to do his calculations with Morton's skull data for *The Mismeasure of Man.* Surely, I thought, in these I would find a clue as to whether Gould or Lewis et al. were truly right about the nature of science. But as I shuffled through the piles of loose legal papers, I found little to aid me to have the final word.[10] Frustrated at my own deficiencies, I wrote a vague discussion of the incident in the form of a lengthy footnote and buried it away. Perhaps for later, I thought.

In the meantime, others more ably came to Gould's defense. Michael Weisberg and Diane Paul published a comment in *PLOS Biology* (upon which Weisberg also elaborated in a piece for *Evolution & Development*) that highlighted several key criticisms of Lewis et al.'s assessment of *The Mismeasure of Man.* They acknowledged that the *PLOS Biology* study did uncover some analytical errors in Gould's work (for instance, his analysis arbitrarily restricted the set of Morton's data under consideration), but they affirmed Gould's claims about racial bias in Morton's cranial study.[11] The debate over the anthropologists' criticisms of Gould continued in various academic journals for the next several years, as scholars in biology, anthropology, and the history and philosophy of science weighed in on whether Morton was a paradigm of racism in science.[12]

I sincerely admire the dedication of my colleagues to vindicating Gould's conclusions, but something vital is missing from this debate. In order to contest Gould's claim that racism has shaped the history of science, the anthropologists who wrote the *PLOS Biology* study *decided to remeasure Morton's skulls*—skulls that Morton had acquired through networks of theft, colonial trade, and grave robbing.[13] They poured shot into the skulls of America's unburied dead and concluded that "biased scientists are inevitable, biased results are not."[14] But why was Morton able to collect the skulls in the first place? Morton owned those skulls because he was at the center of a newly professionalizing network of scientific and medical men in the early American republic.[15] He owned them because he believed that empirical measurement and observation were more reliable methods for understanding racial divisions and race-mixing than Christian scripture.[16] He owned them because he could—because the slave society in which he lived afforded him the ability to be the knower and doer of science rather than its object of study. Even if every single one of Morton's conclusions about the skulls is correct, even if he analyzed the data without any conscious or unconscious racial bias about cranial size and intelligence, we still must contend with the racism, colonialism, and slavery that placed those skulls within his grasp in the first place.

I cannot fault the anthropologists of the *PLOS Biology* study for framing the issue of racism in science in the way that they did. After all, they articulated it precisely as Gould had in *The Mismeasure of Man*. Gould's understanding of scientific racism centered on the critique of biological determinism, the idea that our political life is defined by biological differences among human groups. Whether in the pages of his popular monthly column for *Natural History* magazine, in his correspondence with colleagues, or in *The Mismeasure of Man*, he treated biological determinism as a conservative political ideology.[17] He argued that elites' desire to maintain the inequalities of the status quo—to stay on top of hierarchies of class, gender, or race—drove them to embrace scientific authority to justify their social dominance. He was not alone in this view. Much of the work by figures like Hubbard and Lowe in feminist radical science groups also criticized biological determinism. Feminists sought ways to discredit scientific claims that women were less capable than men at STEM careers or were naturally suited to domesticity. As we will see in this history, they did this initially by highlighting the deficiencies of sociobiology and evolutionary psychology within the context of the natural sciences. These efforts later developed into humanistic critiques that took on the epistemological assumptions of Western science. These scholars used the critique of science to demonstrate the paucity of its claims to objectivity and therefore its lack of authority over the biological capacities of oppressed groups.

For at least a generation, these have been the primary terms of the public debate over the nature of science and its relationship to colonialism, slavery, genocide, and patriarchy. Liberal defenders argue that science is a truthful account of reality. While its use in oppression may be regrettable, they argue, prejudice and violence have been features of humanity from time immemorial and are not especially the fault of science itself.[18] Leftists, though, continue to document the undeniable ways in which social biases have shaped scientific knowledge and practice. Perhaps unintentionally, the terms of this debate have cultivated a broadly held intuition that either science is a truthful account of the world or it is racist and sexist and, furthermore, that the influence of racism and sexism in science primarily functions to denigrate the biological natures of women, Black, Brown, or queer people.

But while biological determinism and social bias are intrinsic to science, they are not the only aspects of its violent history and perhaps not even the most important. If we concentrate too much on how scientific claims are used to preserve social hierarchies, we miss how exploitation and abuse have been inherent to the creation of reliable natural knowledge. Throughout the history of science, individual researchers, scientific schools, institutions, and governments have used the bodies of the most vulnerable to produce masterful epistemologies. It has been in these moments—when science has been most innovative and ambitious, when it has produced cures for diseases, reproductive technologies, astronomical insights, or new statistical techniques—that Native lives have been turned into blood samples, mixed-race women into heredity experiments, enslaved people into raw material for medical dissection, and the cells of a poor Black woman into an immortal cell line that has made possible some of most important medical breakthroughs of the last century.[19]

This last point refers to Henrietta Lacks, who died in 1952 and whose cancer cells were taken without her consent or knowledge or that of her family, then used to produce the first immortalized human cell line. HeLa cells were used to create the polio vaccine, research AIDS, develop cancer treatments, explore genome mapping, and study the effects of radiation and other environmental toxins.[20] The cell line is still widely used in research to this day.[21] It is undeniable that this research has lessened the suffering and extended the lives of countless people in the United States and around the world. But Henrietta's family was largely shut out of these benefits: they received no compensation for the commercialization of the HeLa cell line and struggled for a generation to get access to basic health care. It was only through collaboration with the investigative journalist Rebecca Skloot that the family became aware of the full extent of HeLa's legacy. Skloot established a foundation to create a structure for reparations to Henrietta's descendants and to other families involved in "historic research cases without their knowledge or consent."[22]

The racism in the story of HeLa cells does not lie in their use to make claims about the biological inferiority of Black Americans. It lies in the ability of medical researchers to take the tissue of a Black woman and use it to understand the world better. It lies in the use of a cell line so ubiquitous that few of the scientists who depended on it had any idea of its origins before the story of Henrietta became widely known. This structural racism is so all-encompassing that it functions far beyond the influence of racial animus on the part of a single scientist, lab, or biomedical institution. It would be impossible to count those who have benefitted from the research done on HeLa cells. I am sure that some part of my life that has been directly bettered because of Henrietta's tissue. I am positive that my childbirth experience was safer because of gynecological experiments done on enslaved women.[23] I know that I have had access to hormonal birth control because of the labor of Mexican peasants and because of experimental drug trials conducted by Planned Parenthood on Navajo reservations, at the US-Mexico border, and in Jamaica.[24]

I have benefitted from an unequal and undemocratic science. So many of us have.

The story of Henrietta Lacks is a particularly vivid example of inequality at the core of modern science, but the history of science is filled with these stories. Eugenics, craniometry, scientific racism, sociobiology, and evolutionary psychology are not discredited pseudosciences. They are not the missteps of an otherwise grand and benevolent progress of scientific discovery nor simply pretexts for social prejudice. They are the heart of Western science. They have produced and continue to produce knowledge that enables humans to have greater control over nature, to have a more profound understanding of our own biology, and to study our place in the universe. This power has come through the bodies of countless generations of discarded and forgotten people. It is time that we contend with a simple fact: science can be both true and racist.

Criticizing Science to Build American Democracy

For Gould, scientific racism represented a festering wound in American democracy, an error that scientists would have to address if American science and democracy were to reach their full potential. To Gould's left, a larger group of scientific feminists and antiracists developed a more structural critique of what they regarded as a toxic relationship between scientific authority and political exclusion. The question facing us, then, is this: Is the leftist critique of scientific racism and sexism a necessary or even useful route to building a truly democratic society? Or, conversely, are the left's historical and ongoing appeals for a democratic science partly responsible for the outpouring of conservative xenophobia and ethnonationalism that has blossomed on the right in the United States? By discrediting the Enlightenment foundations of Western democracy, has the Left opened the way for the Right to doubt all facts, reason, and expertise?

The explosion of antiscience and antifact politics during the years of the Donald Trump administration—with attacks on the science of vaccines and the COVID-19 pandemic, on climate change, as well as on data on voter access, immigration, and policing—has amplified a crisis of scientific authority in the United States. Public intellectuals as diverse as Michiko Kakutani and Steven Pinker deplored the falsehoods of the Trump administration as an exacerbation of the overall decline in respect for reason and truth in the West.[25] In their view, doubts about science are at the heart of the Trumpian disregard for facts and rationality. The evidence for this claim is manifold; one can see it in Trump's personal disavowal of climate change science, his flirtations with anti-vax policies, and his championship of dubious COVID-19 treatments. Their claim was supported by his administration's attack on scientific and administrative expertise at every level of government. Liberals continue to argue that a staunch defense of science and reason offers the best, perhaps only hope for the future. The outpouring of this sentiment has fueled countless op-eds and the more concrete actions of the 2017 March for Science. Liberals see that science is under attack and demand its defense.

But other voices warn that science itself is not so innocent. Academics, writers, and activists on the left claim that American science has much to answer for in its history—from the involuntary use of enslaved women in nineteenth-century gynecological experiments to the forced sterilization of Native American women and the ongoing disparities in mortality and morbidity for Brown and Black people in the United States in the midst of the COVID-19 pandemic.[26] Leftists argue that unquestioning deference to science is not the way forward. Instead, they call for a radical upending of the institutions, practices, and even ontologies of science to make space for new voices and new bodies. Only a more liberatory science for the people can ensure justice. To the Left, a critical stance toward science is necessary, even at the risk of undermining public confidence in scientific authority.

There are two issues at the heart of the relationship between science and democracy: first, whether scientific institutions, knowledge, and practices help democracy flourish; and second, whether science itself must be democratic to achieve this goal. As American science boomed after World War II and into the early Cold War, cultural critics were more preoccupied with the first question than with the second. Most scientists and political elites were convinced that the empiricism, critical thinking, and open-mindedness of American science would help the nation win out over Soviet communism.[27] But by the late 1960s this confidence in the ability of science to aid democracy was being undermined by critiques from leftist scholars and activists.[28] Leftist social activism, including that associated with the New Left, targeted establishment science as one of the pillars of the Cold War that could be redesigned for social justice rather than for military power.[29] The ties between scientific research and military spending, an emphasis on rationality and technological power—these features of science were criticized by the counterculture, environmentalists, Marxists, feminists, and anti-Vietnam protestors.[30] The Black Panther Party fought against segregation in medical care and in health care facilities.[31] These concerns over the power of science were taken up by the scientific community itself in the radical science movement.[32] Radical physicists and engineers protested the harm brought on by nuclear and chemical weaponry, and feminist biologists denounced biological theories of gender that impeded the women's movement.[33] In a move that went against long-standing assumptions in Western thought, a diversity of leftist intellectuals criticized science itself for being a source of patriarchy, xenophobia, and other social oppressions.

What united these disparate forms of protest was the belief that the wider public had the right, even the duty, to question professional scientific expertise. Science was simply "too important to be left to scientists."[34] What this meant for Gould in the early days of his career, and for other leftist scientists, was that professional scientists had an ethical responsibility to reflect carefully and publicly on the limits of their expertise. That is, they should acknowledge the political misuses of science but also, as Gould put it, that science was "a social phenomenon, a gutsy, human enterprise, not the work of robots programmed to collect pure information" and not simply an inevitable march "to absolute truth."[35]

Gould's emphasis on the historical biases of science, particularly the relationship between evolutionary thought and racism, put him in direct conflict with a Harvard colleague who held to a more classically liberal view of science as the foundation of an enlightened society. This was the biologist Edward O. Wilson, whose 1975 book *Sociobiology: A New Synthesis* sparked a reaction in feminist and radical science circles that turned Harvard's campus into a hotbed of political controversy. Alongside other texts on the evolution of social behavior, including Richard Dawkins's 1976 *The Selfish Gene,* Wilson's book helped promote a new evolutionary narrative that explained American gender roles as the products of prehistoric adaptations that were encoded in humanity's genes. Gould joined feminist biologists, social scientists, and humanists who excoriated these scientific claims, seeing them as little more than a reactionary response to the women's movement.

A new set of voices entered into conversation with these criticisms of scientific expertise in the early 1980s, when creation science seized national attention during the early years of the rising Christian Right. It brought religion and conservative politics to the forefront of American debates over scientific authority. The terms *creation science* and *scientific creationism* were used by young-earth creationists to describe geological accounts that were consistent with a six-thousand-year existence for the earth.[36] Creationists argued that creationism and evolution were alternative scientific accounts of human origins, and they pushed for laws that would guarantee equal time for creation science in public school classrooms.[37]

Scientific creationism was both a direct and an implicit attack on scientific authority. Creationist literature targeted the evidence, models of explanation, and contradictions in evolutionary science.[38] Writers such as Duane Gish and Henry Morris concentrated especially on the latter, highlighting the disagreements between professional biologists as evidence of the intellectual poverty of evolutionary theory. They often singled out Gould's calls for revisions to Darwinian theory as support for their claims.[39] The very existence of a creationist *science* was an even more significant assault on the mechanisms of credentialing in the scientific community. In the last decades of the twentieth century, creationists set up their own research institutes, scholarly publications, and natural history museums.[40] By mirroring the institutions in mainstream science, creationists laid claim to the heart of scientific legitimacy.

In this period of social upheaval, to many liberal and leftist intellectuals it seemed that religious fanatics had sprung out of the woodwork, elected Ronald Reagan to the presidency, and ushered in a war against science, secularism, and liberal values.[41] The tone of surprised indignation on the part of evolutionists was mirrored among those on the left. And in the case of the evolutionary community, liberals and leftists had a common enemy that they feared more than they did each other. Within evangelical circles, the 1970s and 1980s also took on significance as a moment of great social transformation. They decried the legalization of abortion and the gay rights movement as signs of America's moral degradation.[42] Evangelicals believed they were called upon by these political developments to take a stand for biblical truth and family values.

In the face of these religiously inflected criticisms, liberal definitions of science seemed best equipped to defend the authority of the scientific community. The transcendence of Enlightenment science offered a better rebuttal to creationism than leftist warnings on the political abuses and historical oppressions of science. But leftist engagements with science did not abate after 1980. Across the academy, these critiques developed into new and sophisticated epistemologies in a variety of theoretical fields, including critical race, feminist, postcolonial, neo-Marxist, and queer studies. Often broadly referred to as "postmodernist," these fields translated the social justice commitments of political activism into critical examinations of science, technology, and capitalism. But these criticisms of science were severely complicated by the new political landscape. There were even moments when advocates of science accused leftist critics of being aligned with creationists.[43]

In a 2018 interview with the *New York Times Magazine*, the "post-truth philosopher" Bruno Latour reflected on decades of work by critical theorists and leftists on science, lamenting, "I think we were so happy to develop all this critique because we were so *sure* of the authority of science."[44] But in the twenty-first-century world of "reactionary obscurantism," Latour and the editors of the *Times Magazine* were loath to celebrate the successes of scientific debunking, instead seeing Donald Trump, "a president who invents the facts to suit his mood," as the "culmination of this epistemic rot."[45] Even before the Trump presidency, Latour had worried about the role of critical theory in the growth of antiscience on the right, remarking in 2004 that although "conspiracy theories are an absurd deformation of our own arguments . . . these are our weapons nonetheless."[46] Latour's regrets have shaken the foundations of science studies and questioned the very ethics of denaturalizing science.

Although the conspiracies of American ethnonationalism are extraordinarily real, I cannot mourn with Latour that the critique of science has caused "the light of the Enlightenment [to be] slowly turned off, and some sort of darkness appears to have fallen on campuses."[47] Nor can I agree with Pinker that "the ideals of the Enlightenment are . . . a reason to live."[48] Championing an idealized version of Western science, humanism, and reason leaves no space for most of the bodies in this world. It dismisses the basic humanity and cognitive ability of most people. After all, the attempt to undo science was always more than an exercise in sound epistemology. For the communities who protested medical and scientific authority beginning in the 1960s and 1970s, theirs was a quest for survival. And for a new generation of female, Brown, Black, and queer academics, criticizing science opened the possibility that they might enter the life of the mind. Even so, perhaps it is time to recognize that the critique of science was a beginning, not an end, to the work of justice. There is something more that we must do.

Evolution, Race, and Gender

Narrating these debates in a single frame clarifies many of the important developments within twentieth-century evolutionary theory. We recognize how political and public controversies over gender and religion reverberated within professional debates over the future of evolutionary science—and importantly, over whether Darwinian natural selection could remain the core guiding principle of biology. The history of twentieth-century evolutionary biology cannot be understood without analyzing the twin effects of left- and right-wing political controversies. But beyond this, a history of evolutionary science in this period provides a tangible thread with which to follow the emergence of leftist, liberal, and rightist approaches to science, even as these perspectives went on to shape a range of scientific controversies on topics from climate change to vaccines to infectious diseases.

Evolutionary narratives conceptualize how evolution works—how the process of evolution has shaped human nature, fashioned human variation, and placed humans within the natural world.[49] These narratives are found in the technical papers of professional science, as well as in broadly felt imaginaries and metaphors. Understanding the structure and influence of these narratives requires simultaneously investigating the technical writings of evolution and exploring how evolutionary metaphors have circulated in broader social discourses.[50] Since the early nineteenth century, evolutionary narratives have been intimately linked with how race, gender, and sexuality have been defined, regulated, and legally exploited. During the 1970s and 1980s, a new evolutionary narrative explained contemporary American gender roles as the products of prehistoric adaptations that were encoded in humanity's genes. This framework was at the center of sociobiology during the 1970s and 1980s and continued through the emergence of evolutionary psychology in the early 1990s.

The key argument of the new evolutionary narrative was that natural selection acted for genes rather than for individual organisms. What mattered in the survival of the fittest was not the organism but only specific genetic traits. Because of this view, sociobiologists imagined genes as timeless intermediaries between the present and the past. They came to us after surviving, as Dawkins put it, "in some cases for millions of years, in a highly competitive world."[51] And since humans were believed to have evolved on the African savannah, sociobiologists visualized a prehistoric past in which women gathered food and remained in small family circles, while men hunted large game and had numerous sexual partners.[52] According to sociobiology, the gender dynamic of the African savannah was forever encoded in our genes, and it was the reason why men would continue to find success in the competitive professional world, while women remained more suited to domestic life.[53] Since the past remained within us through the heritage of our genes, sociobiologists argued, the sexual, social, and psychological differences between contemporary American men and women were deeply ingrained.

Feminist scholars excoriated the gendered assumptions of this framework and accused sociobiologists of relying on stereotypes of female sexual passivity rather than on empirical research.[54] Gould's initial criticisms of sociobiology also focused on cultural bias in science, a perspective that inspired his argument in *The Mismeasure of Man* that social prejudice could shape scientific practice. However, his primary assault on sociobiology was against its understanding of the evolutionary process. To Gould, sociobiology was complicit in what he defined as "the adaptationist program" of contemporary evolutionary biology, which viewed natural selection "as the sole directing force" of evolution and allowed organisms to freely "travel an evolutionary path to optimal form in an engineer's sense."[55] And he blamed adaptationism for the pervasive metaphor of progress in evolutionary theory. If evolution was primarily a process of finding better solutions to environmental problems, then more recently evolved organisms were inherently superior. The history of life was thus a ladder of ascending excellence, with humanity at its triumphant apex.

Gould attempted to disrupt this "propensity for ordering complex variation as a gradual ascending scale" throughout his career.[56] His first sally against adaptive explanations was also his most famous. In 1979 he published a paper with Richard Lewontin titled "The Spandrels of San Marco and the Panglossian Paradigm," which sparked debates among evolutionary scientists, philosophers, and humanists that continues to the present day.[57] In it, Gould argued that the existing structure of organisms constrained the free action of natural selection; in other words, sometimes a biological feature was not an engineered solution but the happenstance of history. Later in his career, Gould concentrated more heavily on the refinement of his "hierarchical theory" of evolution.[58] For Gould, a *truly* hierarchical theory of evolution would understand it to operate at different levels (e.g., genes, organisms, species) but would also recognize the processes at these levels as causally independent of one another. Natural selection might create optimal solutions (i.e., adaptations) when it acted upon genes or organisms, but the diversity of life was not simply an extrapolation of this process. The selection of species or catastrophic extinction events might undo or complicate the action of natural selection at the level of organisms. Ultimately, Gould was convinced that evolution was a "copiously branching bush, continually pruned by the grim reaper of extinction, not a ladder of predictable progress."[59]

Although Gould's technical writings contained asides to the connection he made between the metaphor of progress in evolution and historical oppressions, it was in his popular texts that he developed the relationship most thoroughly. There he consistently connected the concept of evolutionary progress with scientific justifications for racism and other social inequalities. In his *Wonderful Life: The Burgess Shale and the Nature of History*, published in 1989, he declared an intellectual war against the "manifestly false iconograph[y]" of evolution as a ladder.[60] And in one of his early columns for *Natural History* magazine, Gould argued that "the fallacious equation of organic evolution with progress continues to have unfortunate consequences," as it encouraged the "rank[ing] of human groups and cultures according to their assumed levels of evolutionary attainment, with (not surprisingly) white Europeans at the top and their conquered colonies at the bottom."[61] Reading these popular texts together with Gould's scientific papers, it is clear that Gould was convinced that transforming the reigning evolutionary paradigm would open the way for political and social equality. And his convictions were partly derived from his family background and his own religious and racial identity. The grandson of Eastern European Jewish immigrants and the son of working-class parents from Queens, Gould was acutely aware that he would have landed near the bottom of nineteenth-century evolutionary hierarchies.[62]

The picture of evolution that emerged from Gould's writings was not one of engineered (either by God or by natural selection) necessity but the result of historical contingency. Showing that the world could have been different made it possible to imagine a new and more just future. Throughout his career, Gould attempted to create a space for this way of understanding evolution—and he wanted this view to transform all evolutionary theory. But Gould began his quest on the eve of creation science's ascendancy into the public consciousness of American political culture. In the face of the political threat of creationism, evolution's defenders distilled their case into the capacity of natural selection to explain the adaptive complexity of life without recourse to the action of God. For Dawkins, for instance, natural selection's ability to fashion adaptations was proof against intentional design and thus evidence against a supernatural designer. His 1986 book *The Blind Watchmaker: Why the Evidence of Evolution Reveals a Universe Without Design* contended that natural selection was the ultimate proof against a theistic worldview. By the mid-1980s, Darwinism involved more than an agreement that natural selection was the primary mechanism of evolutionary change; it was also an assertion that natural selection was *the* objective proof that American creationists were epistemological failures.

In this new political climate, Gould's attempts to diminish the adaptive power of natural selection were barely tolerated by other evolutionary biologists. In fact, the evolutionary biology community interpreted any objections to adaptation as wholesale attacks on the scientific worldview rather than critiques of specific research models. This was how evolution's defenders responded, for instance, when feminists objected to adaptive accounts of the female orgasm in the 1990s.[63] By the end of the century, a new evolutionary narrative, one with the adaptive power of natural selection at its core, had made itself impervious to creationism. By so doing, it resisted ongoing critiques of biological racism and sexism.

Gould saw his attacks on natural selection as actions taken within the scientific community to correct biases in the conduct of science. He did not view them as an assault on the basic authority of science itself. But in the eyes of his more liberal colleagues, particularly as aspects of creationism mirrored Gould's arguments, his criticism of the power of natural selection threatened public confidence in all Western science. His story reveals that the split that opened between leftists and liberals over science and democracy was not solely a matter of public rhetoric. The leftist conviction that science must be critiqued and the liberal insistence that science must be defended went to the very heart of the most technical debates over the future of Darwinian evolutionary theory.

Gould's story is important, in other words, because it shows us that a more democratic science is problematic not simply because of how science is used or applied. Rather, at stake is the very possibility for a shared set of facts and agreement over the reliability of expertise. Moreover, the liberal fear that the public critique of science would undermine its authority was in fact realized in these decades through the explosive influence of modern creationism. Opening science to the people could lead to its downfall, rendering it no more than a single perspective in a host of competing ideologies deplored by liberals, including Marxism, feminism, New Age spirituality, Protestant fundamentalism, and Islam. Democratizing science was not just a pathway to justice; it was also a potential road to conspiracy and anarchy.

Throughout these decades, leftist, liberal, and religious conservative communities understood the nature and purpose of science differently and articulated conflicting views of its relationship to the ethical project of American democracy. Creation scientists and proponents of intelligent design believed that the materialism of evolutionary science encouraged moral relativism, which they blamed for contemporary vices in American culture. Undermining evolution was a critical part of their campaign to return America to traditional family values.[64] Liberal public intellectuals, for their part, argued that science was the necessary foundation of democracy. Humanity's path from a mire of superstition and ignorance into the light of modernity followed the trail of Enlightenment rationality. Science was humanity's best pathway forward and should be defended against cultural doubt in all its forms. Finally, leftist scholars and activists contended that science was complicit in the worst sins of our nation's history. They saw the hand of scientific racism in slavery and the work of eugenics in the poverty and disenfranchisement of generations of immigrants. These evils were so significant that it was incumbent on experts to expose the limits and faults of science to a wider public. When debating empirical matters about evolution—the age of the earth, the provenance of natural selection, or the evidence of the fossil record—each of these groups was committing to a vision of science and the shape and character of American democracy.

Scientific Fears and Democracy

When I first began to study Gould, I wanted to know the answer to a very simple question: how had creationism influenced the fame of evolutionary biologists in the final decades of the twentieth century? In pursuing this question, I became convinced that the public controversy over creationism had impacted technical debates over the character of evolutionary causation and the shape of evolutionary history. But as I read across the disparate popular, technical, and public spaces of these conflicts, I kept finding resonances across histories that had been studied separately—the activism of the civil rights and women's health movements, the science of postsynthesis evolutionary biology, the sociobiology debate over gender, sexuality and race, and the American political realignment that surrounded the Christian Right.

Through all this Gould became my Virgil. It was by following him that I could delve into the technical debates over selection and adaptation and then into the legal disputes over creation science. I could find my way into the radical science and feminist health movements and then delve into the world of evangelical gender discipleship literature. Tracing Gould's career was in some ways an artifact of my own learning, but it remained useful for bringing coherence to the disparate strands of the history of public criticism of science in the United States. In the early days of his career, in the 1970s, Gould was an outspoken leftist who criticized his liberal evolutionary colleagues for their approach to science and politics. Amid the New Left activism of the decade, Gould pushed for a more transparent science that involved the wider public in professional disputes. But during the rightward political swing of the 1980s he became a stalwart defender of academic scientific expertise against creationism. Throughout these political upheavals, he was also a relentlessly ambitious scientist who pushed for a reformation of Darwinian theory that downplayed the role of natural selection in evolution. And so I held on to Gould, at first deploying the language of religiosity to make sense of Gould and his critics—describing him as a heretic who battled against a Darwinian orthodoxy that had been fashioned and deployed in a variety of cultural settings.[65]

But I soon came to realize that I was playing the role of the dispassionate observer. I wrote with the archive as a kind of proof text against which to adjudicate a series of technical and political scientific conflicts. Was Gould or Dawkins right about the level of evolutionary selection? Was sociobiology sexist? Were creationists a threat to professional scientific authority? Certainly, the archive provides a unique lens for contextualizing the various communities who criticized science in this period. But I am not dispassionate. I wrote this history because I want to live in a place where we imagine more fully who can be free and on what terms. I want to understand the power of science and how to wrestle with the temptations of that power. And I want to build a society that has the humility to reckon with its histories of nationalism, colonialism, and racism and to confront how our lives are predicated on the oppression of others. And so I came to see in Gould something else, not a case study or a historical guide but a conversation partner. Gould developed a political vision that was distinctive while also resonating with the wider academic Left's approach to science and political justice. He wrote with a conviction that individual and collective self-reflection on the history of scientific racism could produce a better science and a more equitable society. This book contends with the promise as well as the limitations of that political vision.

Each chapter begins with an episode from Gould's public career, then opens into the wider historical context, and concludes with a reflection on the relationship between the criticism of science and the creation of a democratic society. Although all the chapters move between the technical details of professional evolutionary science and the broader cultural landscape, there are differences in degree. Readers who are primarily invested in the professional history of evolutionary theory in this period may be most interested in the detailed examination of Gould's evolutionary models in chapters 3 and 5 and less concerned, for instance, with the reflections on creationism as a form of populist epistemology at the end of chapter 4. In general, it would be possible to read all the chapters without their conclusions and elide my political musings. However, those readers who are most interested in the political discussions and trust that the archival history supports this could finish this introduction, continue to the openings and closings of all the chapters, and read the epilogue in order to consider the main thrust of my argument that the leftist critique of science is a necessary part of the work of a multiracial American democracy.

Chapter 1 presents Gould as emblematic of many in his generation who sought to bring the ethos of the New Left, particularly perspectives from civil rights and anti-Vietnam activism, into their academic work. Gould's efforts primarily took the form of writing for public audiences, especially on race science for *Natural History* magazine. Importantly, the efforts of leftist scientists like Gould in the radical science movement in the early 1970s largely operated separately from approaches to science in critical race theory and in African American studies.

Nevertheless, Gould's intuition that humanistic accounts of science were key to its democratic possibilities became widespread across the academic Left as the decade wore on. This chapter concludes by meditating on both the promise and the limitations of the critique of professional science as a means of contending with scientific racism.

Chapters 2 and 3 move from these generational beginnings to the first public battle between scientific liberals and leftists: the sociobiology controversy. Chapter 2 examines how the scientists in the Genes and Gender Collective criticized the lack of evidence for sociobiological claims and highlighted the influence of sexism in its authors' depiction of gender. Both Gould and feminist science studies scholars channeled their critiques of sociobiology into historical and philosophical scholarship. This moved the debate away from the immediacy of public protest and institutionalized the leftist critique of science in the critical humanities rather than in the natural sciences. Although Gould was by no means a leader in the emergence of feminist science studies, he was part of a constellation of scholars who sought to transform the university into an institution of justice. This chapter examines the limitations of biological determinism as a framework for understanding historical racism and sexism in science.

Chapter 3 analyzes Gould's criticism of the sociobiological model of evolutionary causation and history. As his criticism developed, he came to argue that sociobiologists' emphasis on natural selection ignored the role of contingency in evolutionary history. This chapter asks whether historical contingency in evolution is important for our attempts to envision democratic futures.

Chapter 4 shifts to the early 1980s and the dramatic effects of creationism and the rise of the Christian Right on the evolutionary science community. In 1981, Gould was an expert witness and a major figure in the publicity surrounding *McLean v. Arkansas*, over the legality of teaching creation science in public school classrooms. During and after the trial, Gould used his fame to defend Darwinism from creationism while simultaneously advocating for his own punctuated equilibria theory of evolutionary change. Crucially, the shock of the trial burdened many evolutionists with a new moral obligation: to be publicly and vocally anticreationist. This chapter reflects on the structural similarities between the populist approach to scientific knowledge in creationism and in leftist health movements and asks whether criticisms from the left and the right of the American political spectrum are ethically equivalent.

Chapter 5 then analyzes how the scientific response to creationism metastasized into a full-on culture war between liberals and feminists. As Darwin emerged as a public icon of secularism, evolutionary advocates distilled the theory of natural selection into a proof against supernatural belief. However, the engineering capacity of selection continued to be at the center of sociobiological and evolutionary psychological accounts of sex difference and gender roles. This chapter takes up the question whether the academic Left should stop its critique of science's capacity to be racist and sexist in the face of religious and reactionary antiscience movements.

The final chapter recounts the transformation of Darwinian science advocacy into the new scientism of the science wars in the 1990s and the New Atheism in the following decade. These developments further entrenched the divisions between liberal and leftist academics. During these decades, Gould attempted to bridge the divides between the sciences and the humanities (including religion) by celebrating the fullness of the Western intellectual tradition. His most famous attempt came in his proposal for *nonoverlapping magisteria*, or NOMA, as a model for achieving harmony between science and religion. But his proposal gained little traction with scientific liberals, who saw only the value of Western science, or with academic leftists, who wanted to unmake the edifice of the project of Western identity itself. This chapter ends by reflecting on whether the struggle for a democratic science will bring about the end of Western secularism and whether such an end is something to be feared. I conclude the book with a claim contained in the epilogue's title, "Why We Can't Wait for a Democratic Science."

This history responds to a deep-seated fear: that if we undo the credibility of science, we will live in chaos and conspiracy. I did not take this concern seriously enough before the fall of 2016. Like Latour, I believed that the authority of science was assured. It was only the week after Trump's election that I learned the etymology of democracy. I had not known that *demos* was the root of both *democracy* and *demagogue*. A demagogue derives his power from holding the demos in thrall, by showing the people that they need not defer to the authority or expertise of the elite. The lines between democracy, populism, and tyranny were thinner than I had imagined.

In response to this fear, those whom I variously refer to as "sociobiologists," "scientific liberals," "liberals," and the "New Atheists" wrote books and op-eds attempting to defend science and liberalism. These scholars and public intellectuals, many though not all of them white men, worried that Western liberalism had been co-opted by the identity politics and oppressive political correctness of the Left.[66] They suggested that the Left was at least partly to blame for the rising influence of the religious nationalism of the Right. But this has not been the only fear at work in American public life. Evangelicals were afraid that scientific materialism had removed God from public life and that the resulting moral relativism was responsible for the increased violence and lawlessness in American society. Leftists, for their part, were most frightened by the history of scientific oppression, seeing the hand of scientific racism in mass incarceration, the work of eugenics in the disenfranchisement of immigrants, and the influence of scientific sexism in the oppression of women. In framing this history through the fears of these communities, I build my own, and I hope my readers', historical compassion, especially for those with whom we might ultimately disagree.

I do not attempt to be objective. Indeed, my feminist training questions the very possibility of such a claim—and regards such claims as too often violent. My aim (at which I do not always succeed) is to write about the communities in this history with kindness and not anger. Writing from anger about the past is a powerful tool; I am often inspired by those whose fury over the atrocities of history offers their readers important moral clarity. But I write to attempt understanding: What did liberals fear the critique of science would bring about? What did evangelicals desire in their quest for a Christian America? Why did leftists believe that reckoning with the violence of history would bring liberation for the future? What were the different visions of scientific epistemology in these three communities, and how did this translate into different visions of the American nation?

But in the end, I am positioned as I am: as a feminist and a decidedly leftist political actor. Although I am no longer an evangelical, I am a Christian. And I am not white but grew up in a San Diego suburb twenty minutes north of Mexico as a daughter of the Filipino diaspora. I explain all of this not only from a long tradition of feminist standpoint theory but also to emphasize that this is a history with an ethical purpose. I tell this story through Gould because in the end I most want to understand why a generation of leftists believed that a different epistemological view of science was a force for social justice. I want to know what problems they believed humanistic accounts of science would solve and why they had faith that denaturalizing science was for the greater good. Through this historical work, I have come to understand that we must engage critically with science, even as doing so carries the real risk of conspiracy and disinformation. But I no longer have faith that the critique of science is enough. It is a beginning, but it is not a resting place. We must do the work of unmaking an old world while also creating a new one.

Being right is not inconsequential; caring about truth or knowing the world better is not folly. I do not disagree with scientific liberals on this point. Nor with evangelicals. But I object to their convictions that cosmology bears its own good fruit, that if we discipline our worldview—either rejecting all supernatural belief or subscribing entirely to a scriptural account of humanity—all else will naturally fall into place. That kind of righteousness makes empathy and love more difficult. And this is, ultimately, also the limit of scientific critique. Its authors also believed that good cosmology would save us, that if we were more right about science than the scientists, a new and just world would be born. But perhaps it is better to live with doubt, even about our own critiques, so that humility is possible. Let us all take the risk of being wrong so that we might focus instead on living generously with one another.

Racism and the Promise of a Democratic Science

> I merely point out that a specific claim purporting to demonstrate a mean genetic deficiency in the intelligence of American blacks rests upon no new facts whatever and can cite no valid data in its support. It is just as likely that blacks have a genetic advantage over whites. And, either way, it doesn't matter a damn. An individual cannot be judged by his group mean.
>
> —*"Racist Arguments and I.Q.,"* 1974

When Lewis Gegner closed his small barbershop in Yellow Springs, Ohio, in 1964, it was a decision that made national headlines. For over a year, Gegner's shop had been the site of weekly sit-ins conducted by students from two historically Black institutions nearby—Wilberforce University and Central State College—as well as the politically reformist Antioch College. Despite his state's antisegregation laws, Gegner had firmly refused to cut Black men's hair in his shop. But when police violence escalated at the growing student protests, Gegner decided to shut his doors rather than integrate his shop.[1]

One of the students who participated in those weekly sit-ins was a young Stephen Jay Gould, who had begun attending Antioch in 1958. Like so many of his classmates, Gould was a student activist; as a member of Students for a Democratic Society and the Student Peace Union, he took part in several protests against US foreign policy and nuclear weapons in Cuba.[2] It was through these circles that Gould became involved in the protests at Gegner's shop. He kept the organizational materials from the sit-ins, including a schematic of the barbershop showing his seat location, as well as a series of photos with him in the corner, watching the events unfold.[3] It was only once Gould had graduated from Antioch and entered a doctoral program at Columbia University in New York that he received news of the shop's closing. He wrote excitedly to his fellow student activists touting the significance of the moment: "It's terribly hard to judge sitting in the midst of SW Ohio, but from a New York standpoint, let me tell you this is the number one national Civil Rights story of the week." Moreover, he argued, the sit-in should not be dismissed as "self-contained" but was "to be viewed as an integral part of the most important national struggle of the century."[4]

And indeed, the civil rights movement forced white Americans to confront the hypocrisies and limitations of American democracy. The segregation, disenfranchisement, and brutalization of Black Americans negated all pretensions to a unique American liberty. As global decolonization movements cast light on America's ongoing race and class oppression, the writings of Black theologians, activists, and intellectuals reimagined a truly liberated American society.[5] But was a multiracial democracy possible? White segregationists' violence cast serious doubt on this possibility. And the ongoing oppression of Black Americans even after the Civil Rights Act of 1964 and the Voting Rights Act of 1965 seriously undermined the liberal hope for an American democracy composed of assimilated and equal individuals.

In the decades that followed, New Left politics were profoundly shaped by the legacies of the civil rights movement and protests against the Vietnam War. In this context, a new set of scientific activist groups abandoned the liberal approach that had guided previous generations of working scientists. Liberalism framed science as a source of neutral facts that could be offered for civic decision-making and relied upon keeping overt political work out of professional scientific organizations. But in what came to be known as the radical science movement, professional science emerged as a key site for political activism. Groups such as Science for the People (SftP) rejected liberal reform in favor of more radical upending of the status quo. Most consequently, these radical science groups extended the ethos of the women's rights movement—"the personal is political"—into an argument that scientific knowledge was shaped by the identity of the scientist. And since most of Western science had been conducted by upper-class European and American men, the movement argued that science in its essentials (not just its misuses) was complicit in historical violence, sexism, racism, and class exploitation. By highlighting this history, scientists who participated in the radical science movement worked for what they believed was a more democratic science that redressed scientific oppression.

The radical science movement was not unique in its leftist politics nor even in its focus on science and medicine. Health activism was a key component of New Left political work—whether the focus on reproductive rights in the women's movement, health care access by the Young Lords Party, or protests against segregation in medicine by the Black Panther Party.[6] What was distinctive about the radical science movement was its conviction that public critique of professional science was a force for justice. And in the early years of the movement, many of these critiques focused on contemporary race science. Gould and other leftist scientists viewed biological explanations of race as the work of reactionary politics and as fuel for segregationist policies. In the first months of his *Natural History* column, Gould devoted considerable energy to dismantling the technical and ideological foundations of contemporary evolutionary and genetic accounts of race difference. There were, however, limitations to this approach. Despite connections between groups like SftP and the Black Panther Party, the primarily white radical science movement did not significantly participate in efforts to build Black political power. Radical science approaches to racial justice would continue to rely on moral persuasion and color-blind liberalism as the best means for racial justice. And this fact had important consequences for the epistemic approach to race in academic scientific critique through the end of the century.

When Gould began writing for *Natural History* magazine in 1974, the monthly column gave him space to model one version of leftist democratic science. Through his persona as a credentialed professional scientist, Gould invited his readers to scrutinize science, particularly with regard to race differences. He translated the politics of protest and activism into a new mode of academic scholarship, one that deliberately opened scientific debate to the scrutiny of public audiences. Examining Gould's trajectory from an undergraduate participating in civil rights protests to a Harvard professor writing for *Natural History* provides an important window into how New Left politics was institutionalized for this generation of American academics. Their aim was to demonstrate that the actual practice of professional science had historically been influenced by racism and sexism. This differed from the liberal stance, which acknowledged that science was sometimes misused to support social prejudices but rejected the notion that scientific knowledge itself was tainted by racism.

The efforts of these leftist academics led to new ways of thinking and talking about science in relation to American democracy. Most importantly, the writings by Gould and others emphasized the possibility that science had been part of the oppressive systems that kept political power out of the hands of women—African Americans, immigrants, and Native Americans. But even this perspective had limitations. The radical science movement's emphasis on racial bias in science sometimes led it to focus too much on the beliefs of individual scientists. This focus on individual belief, moreover, encouraged a broad intuition that better science could be achieved simply through the moral corrections of anti-bias training and self-regulation. In light of this history, how ought we to contend with scientific racism? Are racist and sexist biases in the practice of science our most pressing concern?

We can now build on the insights and work of the radical science movement while also more fully contending with the racism of Western science. Key to this work is a collective recognition that it is in the desire to measure and control that Western science has exacted exploitation and oppression. Our problem is not solely in the *mis*measurement of man but in measurement itself.

Race Science and American Democracy

One of the most widely circulated hypotheses about evolution in the 1960s was the killer ape theory, which emerged from Robert Ardrey's *African Genesis*, Konrad Lorenz's *On Aggression*, and Desmond Morris in *The Naked Ape*. Published in the United States in quick succession in 1966 and 1967, these books promoted a view of human evolution as inherently violent.[7] The enormous popularity of these pessimistic books rapidly shifted American cultural intuitions about the progress of human evolution, even as the country continued to reel from ongoing civil unrest.[8] The legislative gains of the civil rights movement in 1964 and 1965 had not curtailed these conflicts. In fact, the summer of 1964 in Harlem and Philadelphia bore witness to some of the most intense encounters of the decade between Black communities and police forces. The killer ape theory seemed to provide an explanation for the ongoing violence: race conflict was an inevitable, perhaps even necessary part of the progress of human evolution. Ardrey, Lorenz, and Morris, in other words, promoted an evolutionary narrative that contradicted the assimilationist vision of liberal democracy, in which Black Americans could be folded into the full rights of citizenship through legislation, lobbying, and moral persuasion. Instead, their writings suggested that there was biological evidence to support ongoing race segregation of some form.

A second set of scientific theories went even further to claim that America's troubled race relations were the result of the genetic inferiority of Black Americans. One of the most vocal advocates of this view in the mid-1960s was William Shockley, who had garnered fame and prestige after winning the Nobel Prize in Physics for his work in computing in 1956. Shockley vocally opposed desegregation and openly promoted eugenic policies aimed against interracial reproduction. In 1967 Shockley began urging the National Academy of Sciences to study the genetic basis for Black poverty. Shockley cast the academy's initial refusal as an impediment to academic freedom. Although he never succeeded in forcing the academy to sponsor this research, Shockley was nevertheless able to generate a conversation within the scientific community that portrayed the academy's stance as a threat to the free inquiry of science.[9]

This conversation only gained steam after the psychologist Arthur Jensen published his essay "How Much Can We Boost IQ and Scholastic Achievement?" in the *Harvard Educational Review* in 1969. Jensen's argument that the differences in scholastic achievement between white and Black schoolchildren could be the result of heritable differences in IQ—and that it was incumbent upon scientists to study this hypothesis—set off what became known as the race and IQ debate, one of the most widespread academic debates of the mid-twentieth century.[10]

Jensen's argument took aim at compensatory education programs such as Project Head Start, part of Lyndon Johnson's War on Poverty. These programs were intended to address achievement disparities between low- and high-income children and between children of different races by investing in preschool and early childhood education. Jensen opened his attack on Head Start by referencing a 1967 statement from the US Commission on Civil Rights, which had assessed the results of the program. The commission claimed that although the objective had been to "raise the academic achievement of disadvantaged children," by "this standard the program did not show evidence of much success." Given that, as Jensen put it, these "treatments do not cure," it was necessary "to question the basic assumptions" of the program.[11] The key assumption that Jensen wanted to question was that children of different races had the same basic intellectual capacity and that differences in IQ or scholastic achievement were the result of environmental factors.

Jensen roundly rejected the liberal view of the underlying causes of IQ disparities between socioeconomic and racial groups in the United States. The liberal consensus highlighted racism itself as an environmental factor that hampered the progress of Black children; dozens of psychologists had testified to this effect in the landmark 1954 Supreme Court case, *Brown v. Board of Education*.[12] Jensen, by contrast, asserted that science needed to take a "no holds barred" approach to the question whether differences in IQ between white and Black children were based on genetic rather than environmental factors.[13] After considering the evidence to date, Jensen concluded that it was "a not unreasonable hypothesis that genetic factors are strongly implicated in the average Negro-white intelligence difference."[14] Standing on the authority of objective, dispassionate science, Jensen argued that racial inequality in the United States had a biological rather than a social cause. And he contended that any reasonable scientist should at least consider this a viable hypothesis.

Jensen's work, alongside Shockley's campaign and the popularity of the killer ape theory, upended the assumptions that had tied together American science and liberal politics at midcentury. In the midst of the civil rights movement, and following the 1951 statement by the United Nations Educational, Scientific and Cultural Organization (UNESCO) denying the existence of pure races, the scientific community had largely approached racial justice as a social rather than a biological problem. Scientists who adhered to this liberal political framework regarded science as a neutral source of facts that could inform just civic decisions. But Jensen's stance in the race and IQ debate and Shockley's advocacy within the National Academy of Sciences made it clear that scientific neutrality could be leveraged on behalf of segregation as easily as for assimilationist racial reform.

The breakdown of liberal approaches to science and racial justice opened a space for the activist groups of the radical science movement. Although on different terms, these scientists also rejected the liberal reformist stance. As the manifesto document for SftP declared in 1972, "What is needed now is not liberal reform, but a radical attack."[15] In the work of SftP, science was depicted as part of a "ruling class ideology" that was a "self-serving manipulation" to keep "the people" down. The authors of the manifesto cited, for instance, "arguments of biological determinism" used to "help keep Black and other Third World people in lower educational tracks." They singled out Jensen specifically as a part of a ruling class that used science to oppress the already disenfranchised.[16] The SftP 1972 manifesto also revealed the group's approach to social justice through scientific critique. The authors argued that much of the problem lay in the aura of "false mystery" that was perpetuated by "the hold of experts" and subsequently "disseminated by education and the mass media."[17] During this period, SftP became notorious for its disruptive protest tactics at scientific meetings. The Boston chapter, for instance, focused its 1969 efforts on holding events at the annual meeting of the American Association for the Advancement of Science (which they referred to as the AAA$), including a symposium called "The Sorry State of Science."[18] The purpose of the symposium, like that of so many other SftP actions, was to unsettle the accepted protocols of the scientific community.

Over the course of the 1970s, this radical political orientation developed into a set of distinctive scholarly approaches to science and social justice. The beginnings of these new approaches can be seen in a critical exchange between Jensen and the geneticist Richard Lewontin in the *Bulletin of the Atomic Scientists* in the summer of 1970. The *Bulletin* had begun in 1945 as a nontechnical academic outlet where scientists involved in atomic technology could communicate their work to a wider public. At the time of the 1970 exchange, Lewontin was an established and well-respected figure in the field of molecular genetics. He was also a prominent leftist activist, deeply involved in SftP and Science for Vietnam. (Only the year before, Lewontin had been a witness, along with his colleague Richard Levin, to the aftermath of the murder of Black Panther leader Fred Hampton by the Chicago Police.)[19] In a piece titled "Race and Intelligence," Lewontin acknowledged that while Jensen's judicious presentation of his claims had "been disarming to geneticists especially," in fact Jensen's *Harvard Review* piece was "not an objective empirical paper" but instead a "closely reasoned ideological document springing . . . from a deep-seated professionalist bias."[20] Lewontin implied that Jensen's claims to scientific objectivity and dispassionate empiricism masked a racist ideology. According to Lewontin, Jensen promoted a worldview in which the answer to the question "Should all black men be unskilled laborers and all black women clean other women's houses?" would be a resounding yes.[21]

In his criticisms of Jensen, Lewontin employed several tactics that became characteristic of the leftist critique of science. First, his article adopted an implicit view that "genetical" arguments were being deployed against social progress in favor of an oppressive status quo. The view that biological determinism was a hallmark of reactionary politics, whereas environmental accounts of human nature were the purview of social reform, would continue to be a hallmark of leftist critiques of science through the end of the century. Second, he simultaneously attacked Jensen's claims on technical grounds and argued that Jensen's racist ideology biased his scientific work. Lewontin described himself as "a very highly qualified geneticist whose field is the study of genetic variation in natural populations" who "found a few faults" with Jensen's approach to quantitative genetics.[22] Lewontin, in other words, leveraged his own scientific expertise and professional credentials to argue that Jensen's work failed on its technical merits.[23] But Lewontin also claimed that Jensen's bad empiricism was almost beside the point because, as he argued, "Jensen's article is not an objective empirical paper which stands or falls on the correctness of his calculation of heritability."[24] As evidence, Lewontin pointed to the "elitist and competitive world view" in Jensen's rhetoric.[25] He closed his criticism by averring, "Professor Jensen has made it fairly clear to me what sort of society he wants. I oppose him."[26]

These tactics were subtly deployed in Lewontin's critique of Jensen, but all three—associating biological arguments with conservative politics, the use of his own scientific credentials, and deploying humanistic techniques (i.e., the examination of rhetoric) to highlight Jensen's racist bias—became more widely used as leftist engagements with racist and sexist science expanded in the academy. What would change over the course of the decade, however, was the relative balance of their use. Lewontin's criticisms of Jensen spent more time on technical points—owing to Lewontin's qualifications as a researching geneticist—than on close readings of Jensen's racist rhetoric. As the decade wore on, Gould and the leaders of the emerging field of feminist science studies would turn increasingly to humanistic techniques in their critiques of science. And this emphasis on critical humanism would have a significant impact on the institutionalization of the leftist critique of science within the academy.

Jensen's response to Lewontin in the *Bulletin of the Atomic Scientists* embodied a newly developing version of the scientific liberal perspective.[27] Throughout his reply, he characterized Lewontin as a radical ideologue and his own work as nonideological science. For instance, he argued that Lewontin's criticisms of his *Harvard Review* article had an "ad hominem flavor" that stemmed mainly from "ideological reasons . . . not . . . scientific grounds."[28] Furthermore, Jensen argued that he had raised "the question of genetic racial intelligence differences in an obviously non-political scholarly context."[29] Throughout, Jensen called on a set of epistemic values associated with the scientific liberal stance. These included intellectual clarity, rationality, and the willingness to face the cold, hard reality of the natural world. Jensen acknowledged that "there is an understandable reluctance to come to grips scientifically with the problem of race difference in intelligence," but he claimed that the tendency to "blur issues and tolerate a degree of vagueness" must be avoided.[30] This moral positioning was a bid to reset the terms of the debate. Jensen claimed that he was not using science for political ends; rather, he had set aside the shackles of dogma and prejudice for the clarity and reality of science.[31]

At issue in these exchanges was whether it was possible to redress the entrenched racial disparities in American society. Jensen deployed a narrative of scientific integrity and academic freedom on behalf of continued racial segregation and social inequality. Lewontin combined his scientific critique of Jensen with a political commitment to racial egalitarianism. Both men argued that science had an appropriate role to play in a democratic society. What was significant about Lewontin's perspective, however, was his willingness to frame his position explicitly as a moral and political stance. This did not mean that Lewontin or the broader radical science movement advocated epistemological relativism or questioned the possibility of scientific realism. Rather, radical scientists wanted to make it difficult for their profession to pretend that social power did not influence the knowledge it produced.[32]

Within this critique, however, were glimmers of an even more radical shift. Lewontin's introduction of humanistic techniques to demonstrate the influence of racist ideology on Jensen's science opened up new possibilities for how debates over science and society would be conducted in the rest of the decade.

Civil Rights and Natural History

The collective work of the radical science movement was largely focused on political action, that is, campus protests, conference demonstrations, or penning political manifestos. Lewontin's pieces for the *Bulletin of the Atomic Scientists* are an early example of how these radical politics shaped a new scholarly sensibility that manifested within the publishing work of the academy. Although the *Bulletin* was not a peer-reviewed technical journal, it was squarely within the conversation of mainstream American science. Gould's columns for *Natural History* came closer to a new kind of radical science scholarship. His critique of scientific explanations of race difference was especially influential in the academic Left's then developing approach to racial justice.

Gould was born to a secular Jewish family at the start of World War II in the Queens borough of New York City. Both sets of his grandparents had emigrated from Eastern Europe in the early years of the century.[33] His parents wanted to instill in young Stephen a morality based on social equality and to enable him to pursue his passion for evolutionary theory and paleontology. His mother, Eleanor Rosenberg, was an artist who taught Gould to read. His father, Leonard Gould, was an active civil rights campaigner who influenced Stephen to embrace a leftist ethic as he came of age during the social protests of the 1960s. Leonard Gould tirelessly advocated for his son's professional ambitions and continued to write cheerful, supportive letters to his "Dearly Boy" even when Gould was at graduate school and then a professor at Harvard.[34] Later in his career, Gould referred to his father as one of his three lifelong heroes, alongside Charles Darwin and Joe DiMaggio.[35]

Leonard, born in 1915 in New York, was a court reporter for the Queens Borough County Court in the Jamaica neighborhood of Queens.[36] A committed Marxist, he claimed in his writings that society was structured by a series of inequalities and that dealing with the ills of society was not a matter of individual ethics but one of social transformation.[37] Because of his Marxist activism, he was investigated in 1968 by the FBI for possible membership in the Communist Party. Although the FBI concluded that Leonard was not a threat to national security, it did document Leonard's active involvement in a number of "socialist" and "radical" causes, including his support for the civil rights and peace (anti–Vietnam War) movements in the 1950s and early 1960s.[38] During Stephen's teenage years, Leonard served on the executive board of the Queens Borough

National Association for the Advancement of Colored People (NAACP), a role that found him writing for the newsletter and conducting letter campaigns for the local office against racial segregation and discrimination. He also wrote and edited a newspaper called *Vets' Voice*, which pushed for an end to hostilities in Vietnam. For Leonard, these two causes were inextricably linked. In 1965, because of the "ominous widening of the Vietnam War," he resigned from his position with the NAACP because the organization had failed to "forge alliances with the peace movement" and retained "its reliance on the Democratic Party" rather than focusing on "independent political action."[39] Leonard's passionate activism against racism even carried him into a letter campaign that touched directly on themes that would occupy his son Stephen's future professional life.

In the summer of 1964, while Stephen was in graduate school at Columbia University, Leonard sent a series of letters to the editors of *American Anthropologist*, the flagship journal of the American Anthropology Society, addressing the racism in a number of articles that had been published by the British anthropologist Adolph H. Schultz. Over the following years, Leonard would continue to write to the editors of *American Anthropologist*, even involving the anthropologist and popular science writer Ashley Montagu in his campaign. Throughout these exchanges, Leonard complained of the impenetrability of academic science and the impossibility for a layman to criticize the racism of professional anthropology. However, he found hope for the future in the work of his son, at one point writing to Montagu of Stephen's "contribution" to the cause of racial justice: "Me, I've contributed a paleontologist to the fray (a son in graduate study currently)."[40]

Leonard saw in Stephen the possibility for a professional scientist who also cared about the radical change necessary to bring about racial equality in American society. Although in later years Gould would rarely speak directly about his father's activism, he kept the letters from this letter campaign in a packet in his office until he died.[41] Although Gould made clear that his political beliefs were "very different from my father's," this episode nevertheless lends context to the moral ecology of Gould's youth.[42] His father not only was an activist but also saw scientific knowledge and the insensitivity of professional scientists as directly complicit in promoting racism.

In addition to forming the foundation of Gould's family life, leftist values also had a significant role in his undergraduate education at Antioch College, which he attended until 1963. Antioch, a private liberal arts college located in Yellow Springs, Ohio, was considered a bastion of progressive thought, social activism, and free speech. A number of national student activist groups had chapters on

Antioch's campus. Gould was a member and active participant in Students for a Democratic Society (SDS) and the Student Peace Union (SPU). SDS was the iconic New Left organization and the most important student activist group from the mid-1960s through the mid-1970s.[43] Gould attended meetings for these groups, kept literature on racism in society, and threw himself into activism.[44] For example, in 1962, during his junior year, Gould participated in several demonstrations organized by the SPU against American nuclear weapons in Cuba.[45]

While he was an undergraduate, Gould participated in civil rights activism through the Antioch chapter of the Congress on Racial Equality (CORE).[46] CORE had been formed by an interracial group of students in Chicago in 1942. Founded on principles of Christian pacifism and Gandhian noncooperation with evil, CORE pioneered many of the nonviolent direct action tactics against segregation that would be utilized by the wider civil rights movement.[47] CORE organized the 1961 Freedom Rides, which challenged segregation on interstate bus travel, after which the organization concentrated on registering voters and integrating a wide variety of social settings, including retail stores, schools, housing, job sites, supermarkets, and public accommodations.[48]

It was the Antioch chapter of CORE that led the efforts to integrate the Gegner barbershop in the village of Yellow Springs, Ohio. By 1960, CORE activists, alongside the Antioch chapter of the NAACP, had successfully desegregated the other businesses in the village. In response to Gegner's ongoing refusal to serve Black men, CORE activists began picketing the shop and held a sit-in in April 1963. Over the course of that year, the barbershop became a center for civil rights activity; in September a group of faculty and students from Antioch held a memorial march outside the shop for the four African American girls killed in the bombing of the 16th Street Baptist Church in Birmingham, Alabama.[49] Gould's participation in these activities thus brought him into contact with one of the central organizations of the civil rights movement and, most crucially, with its philosophy of political transformation.

According to James Farmer, one of the leading figures of the civil rights movement and the first director of CORE, the organization was founded on three principles. First was the "involvement of the people themselves, that is the rank in file [*sic*]," rather than "relying upon the talented tenth" or "upon the experts and the professionals to do the job." Second was was a "rejection, a repudiation of segregation," and third was an "emphasis upon nonviolent direct action."[50] In dismissing reliance on the "talented tenth," Farmer rejected the strategy of an earlier generation of Black reformists, most notably W. E. B. Du Bois, who had argued that a classical education and moral leadership would allow a small percentage of Black men to lift up the rest of the African American community.[51] Instead, Farmer argued for a more radical populism that would involve all the members of the African American community in a direct, if nonviolent, confrontation with the evils of white supremacy.

This vision, that social change must come through a commitment to democratic action by all the people, rather than the leadership of a select few, was echoed across the politics of the New Left. This included the orientation of the radical science movement, whose insistence on democratic practices within science mirrored CORE's reliance on democratic social activism. Furthermore, Farmer's insistence on a targeted repudiation of racial segregation was echoed in the radical science commitment to outspoken confrontation of racism in science.

Only a few short years separated the counter at Gegner's barbershop and Gould's professorship at Harvard. Soon after his final sit-in, he began graduate school at Columbia University under the advisement of the paleontologist Norman Newell. By the fall of 1967 he had filed his dissertation and joined the faculty in Harvard's geology department.[52]

Gould began his professorship at Harvard during a tumultuous time for junior faculty. By the late 1960s, criticism of the military-industrial complex was increasingly standard on university campuses, even as a portion of science faculty members' salaries came from defense contractors.[53] The Harvard campus, like the campuses of many other universities, was the site of continual political protests and social unrest. A Harvard-Radcliffe chapter of SDS had formed in 1964, largely to protest the escalating draft calls emanating from the Gulf of Tonkin Resolution.[54] The Harvard SDS organized its first antiwar march that spring, prompting President Pusey to comment on "a new and rather disturbing seriousness of tone" in the student agitations against the Vietnam War.[55] Harvard SDS organizers formulated an antidraft program and encouraged students to file as conscientious objectors to the war. Over the next few years, the rhetoric of the organization became increasingly militarized, prompting a physical confrontation between Harvard undergrads and visiting Secretary of Defense Robert McNamara in the fall of 1966.[56] In 1968 a series of national events—including the assassinations of Martin Luther King Jr. and Robert Kennedy—caused the situation to boil over in student communities across the country and around the world. At Harvard, students marched against the university administration, demanding that Harvard end its "blatantly racist policies" and eradicate its ROTC program.[57] This atmosphere of radical politics infused the university's campus culture into the early 1970s.

It was in this context that Gould began writing his column for *Natural History* magazine. When Alan Ternes, an editor at *Natural History*, approached Gould in 1973 about the possibility of writing a regular column, Gould was already familiar with the publication from his work at the magazine's institutional home at the American Museum of Natural History in New York. Biologists in the generation before him had been directly involved with *Natural History*. His graduate adviser, Norman Newell, was a regular contributor, and George Gaylord Simpson, whose book *The Tempo and Mode of Evolution* Gould credited with inspiring his fascination with evolution, had written for the magazine for several decades.[58] Over the next decade, Gould would use this column to advance what he understood to be the central premise of the civil rights movement: that individuals should not be judged by the color of their skin. But while in keeping with the theories of color-blind liberalism he had encountered as a youth, Gould's approach to racial justice would increasingly encounter challenges from both sides of the political spectrum.

As he put it in a letter to Ternes, Gould's original vision for the column was fairly broad, combining insights from evolutionary theory with "my divergent interests in the history and philosophy of science, social and political questions bearing upon scientific issues, and the phenomena of life's history on a grand scale." Gould believed that this combination would "lead to a potpourri of articles," including "straight and quaint pieces of old-fashioned natural history" and "controversial diatribes on the politics of pop ethology." From the column's earliest moments, Gould saw it as a space where he could combine his primary intellectual motivations—revolutionizing evolutionary theory, engaging with the history of science, and expressing his dissatisfaction with what he considered politically dangerous and empirically vacuous scientific theories.[59]

A close look at the early days of Gould's column reveals how distinctly the political sensibilities of the New Left influenced his desire to write the column, as well as the topics about which he chose to write. It also demonstrates how Gould's political vision continued to reflect the imprint of the early CORE platform, even as many leaders in the civil rights movement began to lose faith in nonviolent action in the face of brutal resistance from white Americans. In fact, the story of the Gegner barbershop had ended in the spring of 1964 on a triumphant note—but also a violent one. On March 14, 1964, 550 people held a nonviolent protest in front of the shop, to which police responded with tear gas, water hoses, and the arrests of 108 protestors.[60] Only after a subsequent silent march by 1,500 activists, largely from Wilberforce and Central State Colleges, did Gegner close his shop. But this was only harbinger of the violence that would erupt later that summer. The murder of three CORE workers—Michael Schwerner,

Andrew Goodman, and James Chaney—led many within the organization to abandon nonviolence as the dominant political framework for the civil rights movement. In 1966 Farmer stepped down as the national director. He was replaced by Floyd McKissick, under whose leadership CORE adopted a platform based on Black Power, which focused on political self-determination in the Black community.[61]

Gould had had little direct engagement with these developments; he had not, after all, been present at the teargassing and arrests of protestors in Yellow Springs. Nor does it seem that he engaged frequently in further direct-action protests of racism. In later years he would describe his dislike of "the dirty work of organization" and characterize his relationship with the New Left activism of SftP in distant terms.[62] The result of this, I suggest, is that Gould's own sense of the politics of racial justice was in some way frozen in time. That is, Gould's view of race was influenced by a liberal interpretation of the civil rights movement that emphasized color-blind liberalism. It was an understanding of the problems of racial justice that took its cue from the earlier years of King's writings, such as, for instance, the 1963 "Letter from a Birmingham Jail": "We will win our freedom because the sacred heritage of our nation and the eternal will of God are embodied in our echoing demands."[63] This version of King's legacy emphasized the sacred cause of freedom in the founding of the United States, no matter how violated by the sins of slavery. This message, though strident, still offered the possibility of racial reconciliation within the original framework of the American project, in which white Americans and Black Americans found common ground in a society where children were not "judged by the color of their skin but by the content of their character."[64]

Operating in this mode, Gould opted to combat racism through scientific critique rather than through advocacy or building Black political power. In the first six months of the column's existence, half of his essays focused exclusively on this theme. For instance, in his May 1974 column, "Racist Argument and I.Q.," Gould weighed in on the race and IQ debate on terms established by Jensen's 1969 *Harvard Review* piece. Like Lewontin, Gould characterized these seemingly scientific claims as nothing more than the work of reactionary political agendas. Gould slammed Jensen's assertion that his conclusions were based on data instead of racism. In his *Natural History* column he thundered, "We have never . . . had any hard data on genetically based differences among human groups." This was mere politics, he contended, in which "statements that seem to have the sanction of science" were used "to equate egalitarianism with sentimental hope and emotional blindness."[65] He argued instead that "if current biological determinism in the study of human intelligence rests upon no new facts (actually, no facts at all) then why has it become so popular of late? The answer must be social and political."[66]

In a few years, Gould would expand on his views regarding the historical influence of racism on these kinds of scientific claims in *The Mismeasure of Man*. That text treated the relationship between race and the science of intelligence at greater length and with more historical nuance. But the fundamentals of his message about science and racial justice remained the same throughout his career: the data of science can rescue us from the biases of culture. And Gould tended to depict racial bias as the purview of conservative or "reactionary" political agendas. His essential faith in the facts of science, even while acknowledging the reality of racist bias, would eventually put Gould out of step with more radical epistemologies.

But in the 1970s Gould saw the most important task to be demonstrating that there was no biological support for dividing humans into distinct racial groups and that any arguments to the contrary were based in political, not scientific, realities. Gould articulated this point most clearly in his March 1974 essay, "The Race Problem," in which he contended that the "continued racial classification of *Homo sapiens*" was an "outmoded approach." He therefore rejected "a racial classification of humans for the same reason that I prefer not to divide into subspecies the . . . West Indian land snails," which were the subject of his "own research." Weighing in on the subject with the authority of a credentialed Harvard scientist with expertise in taxonomy, Gould reassured his readers that properly done science revealed that biological race designations were, in fact, unscientific. In Gould's telling, it was the unflinching gaze of science rather than the sentimentality of liberalism that revealed the uselessness of biological race categories. He acknowledged that previous scientists, including the evolutionary geneticists Theodosius Dobzhansky and Grant Bogue, had argued that biological races seemed to be "a fact of nature" or "self-evident." But he argued that these scientists were guilty of a "glaring fallacy": geographic variation, "differences in skin color" being the most obvious, was "self-evident." Gould maintained that the "fact of variability does not require the designation of races" and contended that there were "better ways of studying human differences."[67]

Gould acknowledged that humans vary by geography, but he contended that race designations were made because of political, not biological, necessity. In this, he spoke as a credentialed Harvard scientist rather than as a political leftist. By moving the denial of biological race out of the realm of political debate and into the world of scientific taxonomy, Gould asserted the authority of scientific neutrality. It was a political claim that denied the very fact that it was a political strategy. Gould would further develop this approach in the coming decades, first in contests over nuclear armament and then in his views on the relationship between science and religion. But in the early 1970s Gould deployed this tactic in service of a liberal color-blind view of racial equality.[68]

By the time he was writing for *Natural History*, Gould was working alongside Lewontin in Harvard's biology department. The year before arriving at Harvard, Lewontin had published "The Apportionment of Human Diversity" in the journal *Evolutionary Biology*. Using newly developed techniques in population genetics, Lewontin had demonstrated that most of the variation in human populations was found within local geographic groups and that differences historically attributed to "race" were a minor part of human genetic variability: "Our perception of relatively large differences between human races . . . is indeed a biased perception." Lewontin's argument that race was far less significant than "differences between individuals" offered a biological echo of King's exhortation that we judge each child in America by their character rather than by their membership in a race. Lewontin's paper ended his paper with this pronouncement: "Human racial classification is of no social value and is positively destructive of social and human relations. Since such racial classification is now seen to be of virtually no genetic or taxonomic significance either, no justification can be offered for its continuance."[69] Lewontin's view that biological race was unnecessary and socially harmful mirrored Gould's arguments in *Natural History*. Their work implied that evolutionary scientists best contributed to racial justice by demonstrating insufficient evidence for biological human races.

Black Critiques of Science

Gould's political approach to race science revealed cracks within the Left on the question of race and racism in American society. Gould, like many participants in the radical science movement, approached race with the values of color-blind liberalism. While Gould's writing for *Natural History* translated his undergraduate civil rights experiences into scholarly prose, it also demonstrated his lack of ongoing engagement with leftist Black Power movements.

How did different Black activists and academics view science in relation to the racialized conflicts of American democracy? Fights against the segregation and discrimination of American medicine and health were waged by many Black activist groups. This was the case, for instance, with the Combahee River Collective (CRC), which was started in Boston in 1974 by a group of Black lesbian feminists that included Barbara Smith, Beverly Smith, and Demita Frazier. The women came together initially in response to a series of murders of Black women in Boston. They developed their political philosophy as they reflected on the structural powers that had enabled these murders and, more broadly, on their own experiences of heterosexual harassment in Black Power spaces and the racism of the broader feminist movement.[70] A radical socialist collective, they took a "combined anti-racist and anti-sexist position" that also "addressed . . . heterosexism and economic oppression under capitalism."[71] Developing a novel political vocabulary, they set a foundation for Black feminism and radical activism for decades to come.

Although the CRC is known today primarily as a forerunner of intersectional approaches to social justice, the 1977 Combahee River Collective Statement, as well as interviews with early members, reveals how health and medicine were also integral to their politics. When the collective began, Beverly Smith was working as a contraceptive counselor in Boston. During her time at Boston City Hospital, which served "Third World communities," she recognized how poverty and racism created a lack of access to health care and more generally manifested in bodily experiences of trauma.[72] The CRC statement highlighted health and medicine in the CRC's ongoing work, including in the areas of "sterilization abuse, abortion rights, battered women, rape, and health care," which they carried out on college campuses and at women's conferences.[73] Moreover, the statement specifically highlighted not just the practices of medicine but also scientific epistemology: "As Black women we find any type of biological determinism a particularly dangerous and reactionary basis upon which to build a politic."[74] This orientation resonated with the health activism that had been part of the Black Panther Party's work for almost a decade, activism that targeted not only segregationist practices within medicine but also scientific and medical ideas that denigrated Black bodies.[75] The CRC reframed these concerns through a lens of socialist feminism that highlighted the issues facing "Third World Women," a phrase that signaled not only their solidarity with colonized peoples around the world but their own sense of being a colonized group within the United States.

Black activists and intellectuals were deeply engaged with science and medicine in their political work during this period. However, they did not focus their activism on transforming the structures of professional scientific life. By contrast, radical science groups like SftP viewed the disruption of professional science, especially within the academy, as a key political strategy. Their efforts to protest scientific conferences, compel the American Academy for the Advancement of Science to pass resolutions, or to publish socialist manifestos in scientific journals were all attempts to transform the scientific community itself. Importantly,

SftP did have direct connections with the Black Panther Party; for instance, the Chicago chapter of SftP worked to raise bail money and funds for lawyers for arrested party members.[76] However, the makeup of the radical science movement remained largely white, a demographic reality that influenced its relationship to Black radical politics and the Black lesbian socialism of the Combahee River Collective.[77]

Where Black Power politics had the most influence in the academy was not in the radical science movement but in early articulations of critical race theory. Racial justice activism and scholarship in this new era were profoundly critical of the reformist framework of the civil rights movement. The work of scholars such as Derrick Bell revealed a deep pessimism about the democratic and liberatory capacity of American civic society. The first African American man to receive tenure at the Harvard Law School, in 1971, Bell founded the body of legal criticism that came to be known as critical race theory. Bell's arguments contended that white supremacy was at the heart of the American civic and legal systems and that this white supremacy maintained itself even in moments of seeming progress for racial equality. In an assessment of *Brown v. Board of Education*, widely considered a landmark civil rights case for overturning the precedent of separate but equal established by *Plessy v. Ferguson*, Bell argued that "the interest of Blacks in achieving racial equality will be accommodated only when it converges with the interests of whites."[78] Even in the very act of ending racial segregation, Bell contended, "the decision in *Brown* cannot be understood without some consideration of the decision's value to whites," and "not simply those concerned about the immorality of racial equality."[79] Bell and other critical race theorists did not see racism as a violation of American principles but rather as an outgrowth of the nation's intrinsic white supremacy. A far more radical path than having faith in America's legal tradition would be necessary for racial justice.

There is little evidence that Gould interacted with critical race theory, despite being at Harvard at the same time as Bell, or with Black feminism, despite being in the same city as Barbara and Beverly Smith. The approach to racial justice in the biology department was quite different from Bell's work at the law school. Gould's and Lewontin's commitment to a biological universalism for the human race demonstrated a faith in color-blind political neutrality that no longer appealed to scholars on the cutting edge of work on race. These biologists believed that arguments for racial difference were always animated by conservative racism. They also believed that the essential facts of science would prove that race differences were inconsequential, thereby ending the justification for racist social policies. By contrast, Bell and other critical race theorists were far less concerned with proving biological universalism than with arguing against biological essentialism. They saw race itself as the product of history and racist structures, with little to no connection to biological or material realities.

The distinctions between these approaches demonstrated that leftist academics, even on the same campus, had dramatically different views of even shared political causes. Although Gould's foray into public scholarship was motivated by leftist values, values that were themselves the product of the civil rights movement, his vision of racial justice represented only a portion of leftist approaches to the cause. Moreover, although the radical science movement brought New Left values to the work of science in the academy, there was no consensus even among leftists as to what these values were and how they should shape academic work for social justice.

After the close of the decade, the philosopher and social activist Cornel West would articulate a more profound critique of race and Western science in his 1982 book *Prophesy Deliverance! An Afro-American Revolutionary Christianity*. In it, he argued that the fundamental structures and epistemic values of Western science were entangled with, and productive of, white supremacy. He claimed, for instance, that "the very structure of modern discourse at its inception produced forms of rationality, scientificity, and objectivity as well as aesthetic and cultural ideals which require the constitution of the ideas of white supremacy."[80] This was a more radical critique of the relationship between racism and Western science than Gould had offered in his *Natural History* columns. Gould's stance had been beyond what many scientific liberals would come to argue, namely, that science was objective knowledge that was merely misused or misinterpreted for racist political causes. Gould had argued that racism could and did bias the actual outcomes of science and that scientists needed to be self-consciously reflective of their own personal biases to avoid producing racist science. Through this self-reflection, Gould had contended that a nonracist science was possible. West's view introduced a more troubling possibility: that white supremacy was in the very bones of a supposedly neutral science and that empiricism, objectivity, and rationality were, in fact, part of white supremacy. Science was white supremacy, and white supremacy was science. The kind of analysis that West offered, alongside the growing body of critical race theory, inspired a profound distrust of Western knowledge among leftist academics.

The other significant challenge to the liberal color-blind stance was its eventual adoption by American conservatives. Liberal work for racial justice shifted course in 1965, when President Johnson issued Executive Order 11246, committing the federal government to affirmative action in hiring and employment opportunities. The adoption of affirmative action policies signaled liberal pessimism about the capacity of race-blind policies to overturn two centuries of systemic racism. The conservative backlash to these programs singled out the implications of preference in these laws, arguing that affirmative action, not white society, was racist. With this logic, conservatives claimed a moral high ground, as President Nixon did at the 1972 Republican National Convention, saying: "You do not correct an ancient injustice by committing a new one."[81] Even as critical race theory more pointedly highlighted the essential continuities of white supremacy in American law, conservatives argued that any race-sensitive corrections to historical injustice were themselves racist. The backlash to affirmative action expanded over the 1970s, culminating in the election of Ronald Reagan, who promised that American society would return to a meritocratic free market economy. This promise of political and economic neutrality was itself a kind of negation of racial guilt on the part of Reagan and other conservatives. But however disingenuous, it was a powerful rhetorical tool. By the end of the decade, Gould and Lewontin's biological universalism was better poised to support conservative, rather than leftist, approaches to racial justice in American society.

Moral Persuasion vs. Collective Action

Since at least the beginning of the nuclear age, scientists in the United States had wrestled with the ethics of scientific political action. In those years, scientists had focused on individual morality and liberal reformist politics, both of which were deliberately separated from the research practices and professional structures of academic science. The novelty of the radical science approach was its direct, and public, targeting of scientific research. That is, the combination of science and politics was not new, but public critique—and the public shaming of individual scientists—was.[82] In a new era of race and gender identity politics, the radical science movement represented a major disruption to the previous assumptions of American scientific and political culture.

Even though strands of the liberal approach remained visible within the new radical movement—Gould's own column relied upon both moral persuasion and liberal definitions of race—what was consistently distinctive about leftists was their desire to directly involve a wider public in the professional scientific debate. In this light, Gould's public scholarship can be understood as one way of democratizing and including nonexperts in the work of science. The correspondence between Gould and his editors at *Natural History* reveals a mutual desire to make science relatable to a broad readership. Alan Ternes especially pushed Gould to consider the needs and interests of the lay reader. And the fan mail Gould received from readers of his column gave him a direct encounter with people from many backgrounds, most of whom lived and worked outside the sciences and even the academy. Through his public writing, Gould let everyday citizens into the inner workings of the world of Harvard biology.

Was public scholarship an adequate way to satisfy the radical science movement's desire to democratize science? One way to answer this question is to reflect on the column's relationship to other publication venues in the world of Boston leftist activism in the 1970s. In 1970 SftP began publishing *Science for the People* magazine as an outlet for the activist organization's perspective on science and politics. In the magazine's first decade, issues covered leftist topics such as the ethics of war research; the connection between food, agriculture, and American imperialism; and women and science. Gould was a member of the advisory board, but he did not write regularly for *Science for the People.*[83] And even though Gould's archive reveals the connection between his leftist politics and the inception of his *Natural History* column, "This View of Life" was not consistently regarded as a radical science effort. To the contrary, it was a regular feature in a magazine founded in 1900 that benefitted from the institutional prestige of its sponsoring organization, the American Museum of Natural History. And while the column may have begun with a set of distinctive New Left principles, by century's end its most significant legacy was Gould's considerable academic celebrity.

Ultimately, Gould wrote about evolution in his *Natural History* column to express his distinctive evolutionary perspective on life.[84] His own understanding of evolution revealed to him "the abandonment of the hope that we might read a meaning for our lives passively in nature," and that we were instead "compel[led]. . . to seek answers within ourselves."[85] Even so, Gould's column never challenged the essential hierarchy between himself and his readers. Gould remained the expert; although he often adopted a confiding tone, readers were never presumed to be part of the process of making science. Popular science writing was a way of inviting nonscientists to learn about the details of the scientific process, but it did not include everyday citizens in the shaping of research questions or agendas. There were limits to its democratic potential.

In its early days, Gould's audience was made up primarily of educated professionals. His column did not teach basic science facts to the untrained reader but rather recounted stories of scientific principles and the history of science to a highly literate readership. The essays, with their myriad literary references (often bordering on the esoteric), were not for every audience. For instance, in the opening to a 1975 column, "The Child as Man's Real Father," Gould referenced Ponce de

Leon, Chinese alchemists, Faust, Wordsworth, and Aldous Huxley within the first six sentences.[86] His writing presumed a shared, highly literate, cultural background with his readers. When writing of Wordsworth, Gould referenced the poet's "famous ode," assuming that his readers would immediately recognize the quotation "splendor in the grass, of glory in the flower" from the English poet's *Imitations of Immortality* (1807). This particular column was Gould's reflection on the importance of neoteny in human evolution. He ended the essay not by quoting the latest scientific journals on the subject but instead with quotations from John Locke and Alexander Pope.[87] Gould's writing was better at connecting the reader well versed in the humanities with the sciences than it was at providing a general education in scientific facts.[88]

Gould's column would reach a wider audience once it was anthologized in a published volume. As Gould's book editor Edwin Barber would later relate, he was reading old issues of *Natural History* magazine in the New York Public Library when he ran across Gould's monthly column.[89] He contacted Gould, demanding to know why Gould didn't have a book in print. Barber was an editor for W. W. Norton, the largest independently owned publishing company in the country. Norton prided itself on releasing both trade and text titles that were meant "not for a single season but for the years."[90] Gould published extensively with Norton, including the first seven anthologies of his column, two popular science books, and two collaborations with the photographer Rosamond Purcell.[91] The book anthologies created an experience of Gould's ideas that was tidier and more manageable than the columns. The first anthology, *Ever Since Darwin* (1977), did not make Gould an overnight literary sensation, but it did provide an important venue for his nonfiction writing for the next decade.[92]

Although from different regions and varying readerships, the vast majority of the reviews of *Ever Since Darwin* settled on a central message about Gould: here was a professional scientist—concerned with the esoteric, even dull—who had the "natural" gift to write like an angel. Another strategy for explaining Gould's place in the world of science writers was to couple reviews of *Ever Since Darwin* with pieces on other, already well-known scientific popularizers. The *Chicago Sun Times* combined their review of *Ever Since Darwin* with a review of Richard Leakey's *Origins*.[93] This was a signal to potential readers that provided context for Gould and suggested what sort of reader might be interested in his work. Academic journals also reviewed the book, though these reviews were less personable and tended to focus on Gould's professional credentials. One periodical, *Systematic Zoology*, the bimonthly journal of the Society of Systematic Biologists, handled Gould's multifarious identity by listing his disciplinary training: "Stephen Jay

Gould is a very impressive figure. Though only 36, he has produced numerous books and papers. Though formally a professor of geology, he is well versed in biology and the history of science."[94] The academic reviews rarely marveled at Gould's writing talent.

Gould was not a household name at this point, but he was becoming a prominent academic with greater than average public recognition. The hardcover edition of the book sold almost nine thousand copies.[95] *Ever Since Darwin* introduced Gould to a literary sphere that reached larger audiences than *Natural History*. The reviews of the book provided an opportunity for Gould to feature at least once in almost every daily and weekly periodical in the country, from the *New York Times* to the *American Socialist*.[96] By the publication of *A Panda's Thumb* in 1980, Gould had begun to develop a niche for himself as a popularizer perhaps akin to the astronomer and television personality Carl Sagan but unique for his insight into evolutionary theory and his continued original scientific research.

By the mid-1980s, Gould's books and television and magazine appearances had begun to connect him to an ever-expanding circle of notable writers, thinkers, and media personalities. His correspondence reveals that many of these contacts came through his own initiative. For instance, in 1985 he sent a copy of his fourth *Natural History* essay collection to Kurt Vonnegut and received this pleasant reply from the novelist: "So you have surely imagined at least two truths about me: that you were always on my mind when I wrote Galapagos, and that your good opinion would please and comfort me."[97] Gould sent copies of his works not just to other well-known authors but also to former presidents. After Gould received a complimentary letter from President Jimmy Carter on his popular book *Wonderful Life: The Burgess Shale and the Nature of History* in the winter of 1989, the two struck up a friendly correspondence.

Thus, in the late 1970s and early 1980s Gould's *Natural History* column entered him into a world outside both professional science academia and political activism. He was newly connected to a landscape of publishers, filmmakers, broadcasters, editors, and producers. By the 1990s he had attained true celebrity. In the final years of Gould's career, the first world to which he had belonged, the scientific circle, began to occupy less of his time and energy. This is an empirical observation, not a subjective evaluation. It is clear from a survey of Gould's correspondence in the late 1990s that he could no longer claim, as he had in a letter from 1979, that he "[had written] to several biologists around the country just as a matter of course over the past week."[98] Of course, Gould maintained his place in biological circles. He continued to teach at Harvard at least once a week until his final illness, and *The Structure of Evolutionary Theory* (2003) captured much of his attention in his last decade of work. However, Gould also became increasingly part of a literary set, writing more and more reviews for the *New York Review of Books* and appearing as a talking head on *Charlie Rose* and on *CNN Talkback Live*. He was an intellectual luminary who perhaps only happened to be an evolutionary biologist.

Even through this transformation, the column retained a distinctive moral perspective. Gould's title for his column, "This View of Life," celebrated the unexpected and unpredictable side of the natural world in which Gould asserted true moral freedom existed. If the inequalities he saw in American society were quirks of history, rather than part of a creator's plan or the result of evolutionary necessity, then it was possible to eradicate those injustices. In his version of Darwinism, Gould found hope that human evolution had "given us flexibility, not a rigid social structure ordained by natural selection."[99] Many during his lifetime and since have found inspiration in Gould's approach to evolutionary theory and science. Many others have considered him a talented writer who nevertheless built his scientific career on self-aggrandizement. But he was just a man whose desire for power and prestige was at times his greatest flaw. There was a deep irony in Gould's relentless conviction that academic life could be used to address social and historical inequalities, even while he used his considerable intellectual powers to advance his own fame.

How to Confront Racism in Science?

Writing in the early years of Barack Obama's presidency, the legal scholar and author Michelle Alexander worried lest Americans mistakenly conclude that the election of the first Black president meant that the era of American racism was over. In her 2010 book *The New Jim Crow: Mass Incarceration in the Age of Colorblindness*, Alexander argued that the new American "postracialism" was a facade behind which the dramatic rise in mass incarceration of minoritized racial groups had in fact built "a system of social control unparalleled in world history."[100] Beginning in the 1980s with the War on Drugs, the United States imprisoned more of its racial minorities than any other country, incarcerating a larger percentage of its Black citizens than South Africa had done at the height of apartheid.[101] Alexander called this social system a "new Jim Crow" for good reason: a person convicted of a felony in the United States can be stripped of most of the rights of citizenship, including the right to vote or serve on a jury, as well as denied equal access to housing, employment, and other public benefits.[102] Alexander argued that far from ushering in a "benevolent society," Obama's presidency would blind many to the industrial scale of American racial violence and oppression.

When Alexander returned to these themes ten years later in a piece for the *New York Times*, she no longer believed it necessary to persuade her audience that racism was at the heart of American political life. Donald Trump's blatant nationalism fueled the politics of white supremacy against Black communities, Mexican immigrants, and other minoritized racial groups.[103] But even as Trumpian white nationalism took center stage, Alexander argued, the structural racism of American social institutions was not the overnight work of a single president. After all, the scale of prison complexes and immigrant detention centers had escalated at the turn of the twenty-first century, having been steadily built by both Democratic and Republican administrations. What had changed after Trump's election in 2016 was the death of the postracial narrative of American social progress. The political clarity of the Black Lives Matter movement has forced a national debate on the history of American racism; in this context historians and activists have revisited the civil rights era on different terms than the static memorialization of the last century.[104] Rather than regarding the civil rights movement as a set of settled triumphs in an inevitable march toward social justice, scholars enter into a dialogue with this history over a set of still open and as yet unsolved political dilemmas as to whether it will be possible, in Alexander's words, "to birth a truly inclusive, egalitarian democracy."[105]

Similarly, we can engage with the history of the radical science movement with the same openness. Rather than regarding the critique developed by Gould, Lewontin, and others as the final word, we may instead see these as a set of approaches to the (also yet unsolved) political dilemmas over racism in science. The political reassessments of the civil rights era offer an opportunity to reflect on the critique of scientific racism, as well as to develop novel vocabularies to account for racism in science.

Consider, for instance, a volume from 2018 titled *Fifty Years Since MLK*, which assesses what King's legacy "can and cannot do for activism today." King emerges from the essays in this volume not as the "frozen dreamer" but as a dynamic political thinker of both promise and limitation. Of relevance to the themes of this chapter is a piece by the African American studies scholar, historian, and writer Keeanga-Yamahtta Taylor entitled, "The Pivot to Class."[106] In his mature political thought, Taylor argues, King was forced to reckon with the insufficiency of nonviolent civil disobedience to address the type of racial discrimination faced by Black Americans outside the South. The tactics that had proved so successful in dismantling the explicit racism of Jim Crow laws in the US South were ineffective in the ghettos of northern and western cities, where, as Taylor puts it, it was the "insidious but obscured actions of the real estate broker, the banker, the employer, the police officers, and other agents that maintained racial inequality." King struggled to highlight (and therefore shame) the racism of social institutions that claimed to be color-blind.[107] By the end of his life, King had moved away from the conviction that white America could be persuaded to end racial oppression simply by reminding it of the principles of freedom on which the nation had been founded. Rather, in Taylor's telling, King moved toward an anticapitalist and anti-imperial understanding of American history, arguing in a posthumously published essay that only by facing its "interrelated flaws—racism, poverty, militarism and materialism" could America embark on the "radical reconstruction of society itself."[108]

Bringing Taylor's insights on King to a reading of Gould clarifies how Gould's critiques of scientific racism operated with the same political logic as King's earlier nonviolent direct action. (This is perhaps unsurprising, given that Gould's own experiences with the civil rights movement were influenced by a parallel era of CORE activism.) Gould's writing on racism in science for *Natural History* also relied on a kind of direct confrontation of the racial bias that might influence scientific practice. His development of the idea in *The Mismeasure of Man* suggests that he believed that this bias was best addressed through collective self-reflection, that is, through invoking shame over past racism to develop a resolution to shed that racism in the future.

But if we understand science and medicine in a more capacious sense, as something more akin to and linked with prisons, policing, housing, and schooling, then we must change tactics in much the same manner as King. Here again we can look to scholars revisiting the civil rights movement as inspiration for ongoing political engagement. One such scholar is Nishani Frazier, a historian, scholar, and daughter of Cleveland CORE activists. In her 2017 book *Harambee City: The Congress of Racial Equality in Cleveland and the Rise of Black Power Populism* she advocates for the more nuanced picture of community and political institution building that were integral to Black Power populism. Frazier argues that Black self-help—the creation of Black businesses or the fight to open Black hospitals—created social realities within which Black life could exist.[109] Frazier's analysis points us to a different set of legacies from the era of leftist protest and activism. When we consider the efforts by the Black Panther Party, the Young Lords, or the CRC to create health care access for Black and Brown communities, instead of the writings of the New Left, our attention is directed toward the ongoing work by movements that not only identified racism in science but also sought to build new institutional spaces in science and medicine for Black and Brown life to flourish.

Science can also play a role in developing a vocabulary for describing structural racism. This is the case, for instance, in the critical scientific work being done by Terence Keel and his interdisciplinary BioCritical Studies Lab (BCS) at the University of California, Los Angeles. In a report released to the press in the summer of 2022, the BCS research team claimed that the Los Angeles County jail had deliberately misclassified the deaths of a number of Black and Latino men. The team came to this conclusion after evaluating autopsies of deaths in the jail and finding that in a majority of deaths the inmates had suffered "physical violence" even though the cause of death was classified as undetermined or due to natural causes. Racial justice was central to the BCS lab's public release of the report. As Keel stated in an interview with the City News Service of Los Angeles, "Tell us the truth about how our sons are dying in jail. . . . These people are dying in darkness."[110] Importantly, a key component of the lab's Coroner Reports Project is a public-facing data science project done in collaboration with Dignity and Power Now, a Los Angeles–based organization founded in 2012 with the mission to build dignity and power for incarcerated people and their communities.[111] This effort is not primarily directed toward identifying racial bias in science but instead seeks to conduct empirical community-based research that fights against the racism of the prison system.

The intersection of racism and technoscience is also a persistent theme in the writings of the Black feminist STS scholar Ruha Benjamin. For instance, Benjamin's 2019 book *Race After Technology: Abolitionist Tools for the New Jim Coda* (whose subtitle is a deliberate play on the title of Alexander's *New Jim Crow*) theorizes how digital technologies create new forms of discrimination even as they obscure the ways in which they do so. It is in the very desire for "objectivity, efficiency, profitability and progress," she argues, that digital technologies deepen racial discrimination, and it is because of this focus that they are able to appear "neutral or benevolent."[112] Benjamin rejects the portrayal of racism as a social prejudice that may—or may not—infect scientific practice. Instead, she imagines the category of race as a kind of tool, as a technology, as a science, and thus much too connected to Western technoscience to be dealt with by direct confrontation with social bias. It is not possible to imagine a science that is free of racial bias and utilized in a truly neutral manner.

One of Benjamin's most striking arguments is that the way forward to "alternatives to the racist status quo" is to "empl[oy] speculative methods." Benjamin points to ways of knowing and being outside Western science, including spiritual techniques from African diasporic religious traditions. In these invocations, Benjamin suggests that one way to end the technology of race and move toward antiracist futures is through the "everyday spiritual technologies" of "the Black feminist approach to kinship."[113] In other words, we only get free of racism in science when we allow ourselves to look at the world from a perspective free from Western science.

Freeing ourselves from Western science in this way moves us far from the vision of Lewontin, Gould, and other scientists who took part in the radical critique of scientific racism during the 1970s. Particularly after creationism took the national stage, leftist critique did not look to religion as a potential resource in the fight against scientific racism. But in exploring the limits of Gould's perspective on scientific racism, I do not mean to set it aside as a bankrupt or inert philosophy. I believe that examining the details of his vision of political change makes use better able to understand what the critique of racial bias can—and cannot—offer to us if we are committed to that "truly inclusive, egalitarian democracy." What is important here is to recognize how our theories and historical narratives either create or foreclose different political possibilities.

Before we can further explore these possibilities, we must turn to a set of issues that haunts the critique of scientific racism, that is, the historical intersections between race, gender, and sexuality. Thus far, I have addressed these intersections implicitly, but we must now consider how the women's movement, in particular the women's health movement, shaped the radical science critique of scientific oppression. The emergence of a distinctly feminist critique of science sharpened the leftist intuition that science could only participate in social equality by opening itself up to scrutiny by marginalized groups in American society. In the very power of science lay its fallibility and its need for public accountability. But even as feminists institutionalized this critical sensibility across the academic Left, they provoked a reaction on the part of scientists who desired to defend Western science as the authoritative foundation of American democracy. And this scientific liberalism was rebirthed in the midst of a public controversy over a new theory of human social evolution: sociobiology.

Sociobiology and Gender Trouble

> The issue of biological determinism is not an abstract matter to be debated within academic cloisters. These ideas have important consequences . . . the most immediate impact will be felt as male privilege girds its loins to battle a growing women's movement.
>
> —*"The Nonscience of Human Nature,"* 1974

Gould's column for *Natural History* magazine began as a way to balance the political convictions of his civil rights experiences with his desire to revolutionize evolutionary theory. As his career soared to new heights in later decades, his professional ambitions eventually eclipsed his leftist politics. But in the late 1970s he was still using the column to address contemporary debates over science and politics. In the spring of 1976 he decided to weigh in on a controversy close to home with a column titled "Biological Potential vs. Biological Determinism," which joined in the leftist criticism of Edward O. Wilson's 1975 book *Sociobiology: The New Synthesis.*[1]

By then he and Wilson had been colleagues in Harvard's biology department for several years. At first glance, Wilson's book might not have appeared to be the most likely candidate to spark leftist outrage. It was a long academic volume that synthesized empirical work on a host of animal taxa with the aim of clarifying a new program for the evolutionary study of social behavior. Wilson was convinced that the qualities of social life—e.g., aggression, cooperation, and hierarchies—were as much a product of natural selection as were physical traits. And in what would become an infamous last chapter, he extended this argument to the study of human societies.[2] The book was far more empirically grounded in its treatment of human evolution than the popular works of Robert Ardrey, Konrad Lorenz, and Desmond Morris, which had fed into narratives of inevitable race war at the height of civil rights activism.[3] Nevertheless, *Sociobiology* was at the heart of the most consequential debate between the leftist and liberal perspectives on science and American democracy of the era.

Wilson's writings became a flash point as a new set of evolutionary models of sex difference clashed with the political demands of an intense phase of the American women's movement. New legal triumphs that guaranteed the right to contraception for married couples, the right to abortion, and protections against sex-based discrimination were counterbalanced by a ferociously energetic conservative Christian movement that fought against the Equal Rights Amendment and any possibility for changing women's place in American society. Even as women across the country reimagined their roles at home, at work, and at church—and pushed for the legal protections to do so—reactionary politics continually insisted on limiting what women could do and be.

It was in the midst of this political tumult that Wilson's book (alongside other texts on the evolution of social behavior, including Richard Dawkins's 1976 *The Selfish Gene*) promoted a new evolutionary narrative that claimed that contemporary American gender roles were the products of prehistoric adaptations encoded in humanity's genes.[4] Sociobiologists like Wilson and Dawkins envisioned a prehistoric past in which women gathered food and lived in family camps, while men went out to hunt and seek new sexual partners. In subsequent decades, scientists and nonscientists alike would deploy this narrative in both scientific and popular settings to rationalize gender disparities in STEM fields and the workplace and to naturalize rape.[5] Gould's criticism of Wilson was joined by critiques developed by other leftists from the sciences and the humanities, who viewed sociobiology as reactionary politics rather than sound science. And the sustained protest against the sexism of sociobiology over the next two decades would be led by the leaders of feminist science collectives, including Ruth Hubbard, a biologist at Harvard, and Ethel Tobach, a psychologist at the American Museum of Natural History.

Before sending his column on sociobiology to *Natural History* for publication, Gould sent a draft of it to Wilson. Wilson's outraged reply and the subsequent exchange between the two men reveals far more than just the contours of their personal animosity. As expressed in his letters to Gould and in later publications, Wilson had a more classically liberal view of science's proper role in American democracy. Liberals view science as truthful knowledge that serves as a foundation for an enlightened society to guarantee equality and enact rational governance. Thus, they consider science essential for democracy, but they do not prioritize a democratic approach to the actual practice of science. As liberals see it, even when science is only done and understood by a few elite white men, the reliability of its knowledge of the natural world enables it to be the foundation of an equitable society.

This understanding of science and democracy was unacceptable to Gould, as well as to other leftists in the radical and feminist science circles that protested Wilson's book. Although their understanding of science for the people was by no means consistent, members of these movements shared a conviction that the elitism of science impeded its capacity to support democracy. For leftists, the inclusion of women and minoritized racial groups in the professional practice of science was essential if science was to contribute to a progressive society. Wilson, for his part, characterized the attacks by Gould and others in what became known as the Sociobiology Study Group (SSG) as an attempt to restrict the freedom of scientific research and a worrisome sign of intellectual censorship.

By the end of the century, many public scientific liberals would castigate both Gould's historical accounts of scientific racism and the feminist accounts of gender bias in science as "antiscientific." But the history of this late 1970s moment reveals that neither Gould nor feminist scientists saw their criticisms of sociobiology as antiscience. In fact, they understood the debate to be a conversation within the scientific community about the evidence for a new model within evolutionary science. They believed that a better science, one that acknowledged the pitfalls of gender and racial bias, could be achieved through collective self-reflection on the motivations and practices of scientific work. And this better science could, in turn, be used to combat what these leftist academics feared were reactive and oppressive political actions. Their willingness to address the role of social influence in science and to publicly criticize current scientific research, however, set the stage for a new cultural divide. By the end of the century, sociobiology had claimed the mantle of scientific authority on human sexuality. And feminist and other leftist academics struggled to stave off accusations that their approach to scientific knowledge was itself antiscientific.

A Study Group Criticizes Sociobiology

On a cool, clear evening in the spring of 1977 a group of scientists gathered at a public library in Cambridge, Massachusetts, to hold a community teach-in about sexism in science. The hosts distributed workshop discussion starters, all of which focused on the use of genetic determinism to justify conventional gender roles. Participants were instructed to ask such questions as, "How would you go about finding out if male and female roles are biological?" or "How do you distinguish between scientific bullshit and scientific wisdom?"[6]

The academics who spoke about "scientific bullshit" were members of a newly formed radical science organization, the Sociobiology Study Group. The group was started by Lewontin and two other Boston-area scientists, Jonathan Beckwith and Freda Salzman, in response to the publication of Wilson's book. Beckwith, a microbiologist at Harvard, pursued science and activism as an integral project, in particular fighting against cultural narratives that attempted to link genetics and criminal behavior.[7] Salzman, a physicist, found her way into radical science activism after experiencing structural sexism at the University of Massachusetts.[8] When the group began asking other faculty and students in the Boston area to join them, Gould was among the first to say yes. He began attending meetings early in the summer of 1976.[9]

As an affiliate of the Boston chapter of Science for the People (SftP), the SSG was embedded in the wider radical science movement. The group brought the movement's tactics of public protest and professional disruption to their criticisms of Wilson. For the SSG, one of the most alarming features of Wilson's book was the perception of legitimacy associated with Wilson's standing as a Harvard-credentialed scientist.[10] The tipping point for the group came in November 1975, when the *New York Review of Books* carried a review of Wilson's book by the British developmental biologist and paleontologist C. H. Waddington that seemed to suggest that it was a "very good and very important book."[11] In response, the SSG drafted a letter asserting that Wilson's theory had "no scientific support" and upheld a vision of the world "similar to the world which E. O. Wilson inhabits." The letter concluded that Wilson had joined "the long parade of biological determinists whose work has served to buttress the institutions of their society by exonerating them from responsibility."[12] Gould helped draft the letter and was one of its signatories (though Beckwith later affirmed that he added the more controversial phrases to the document, which likened Wilson's work to "eugenic policies" and "gas chambers in Nazi Germany").[13] The publication of the SSG letter in a later issue of the *New York Review of Books* launched a series of high-profile public criticisms of Wilson by the group, including protests in his classroom, publications in professional and popular venues, and other actions in the Boston area.[14]

A theme of the SSG's critiques was that *Sociobiology* legitimated a cultural confidence that sexual stereotypes were in fact genetically determined. Wilson's text depicted contemporary behaviors, including sex differences, as the epiphenomena of ancient genetic programming. For instance, Wilson asserted that "the populace of an American industrial city, no less than a band of hunter-gatherers in the Australian desert is organized around [the nuclear family]." And he argued that within that nuclear family social roles had remained constant over time:

"Women and children remain in the residential area while men forage for game or its symbolic equivalent in the form of barter or money."[15] Wilson's decision to relate the prehistoric past to contemporary America through reference to hunter-gatherers was not itself a departure from a long tradition of using contemporary "primitive" societies as a stand-in for the earlier evolutionary states of "civilized" cultures. What was unusual, however, was Wilson's claim that the evolutionary past (i.e., the primitive state) still controlled present-day "civilized" society.[16] According to *Sociobiology*, the past remains with us through the heritage of genes, the implication being that sexual, social, and psychological differences between contemporary men and women are deeply ingrained.

The SSG's initial letter to the *New York Review of Books* set off a larger media storm, generating headlines over the next several years that recounted the conflict between Wilson and his critics.[17] Although Gould attended several of the gatherings and continued to receive the minutes of the group's meetings, he was by no means its most politically active member.[18] Nevertheless, popular media accounts of the controversy tended to focus on the great interest generated by, as the *Washington Post* journalist Myra MacPherson explained in a letter to a member of the SSG, "the very fact that Lewontin and Gould et al are colleagues of Wilson . . . having such opposite views . . . and that a strained relationship exists."[19] This overshadowing of the rest of the group created frustration among many of its members, who, after all, desired to forge an anti-elitist and distinctly feminist approach to activist work.[20]

Decades afterward, Wilson claimed that he had been a "political naif" with barely any knowledge about the academic Left before the criticisms of the SSG.[21] This claim supported a narrative that Wilson cultivated from this period onward that contrasted his dedication to the dispassionate objectivity of scientific research with his critics' ideological bias. It was certainly the case that members of the SSG were more likely to frame their criticisms of Wilson as contributing to a larger political struggle for social justice.[22] Many of them, like Gould, came to their protests of sociobiology having earlier participated in the nonviolent activism of the civil rights movement.[23] As we have seen, a crucial component of that movement's theory of political change was the need to engage in direct-action confrontations with racism. Many of the actions taken by the SSG can be understood as mirroring this same political logic, but for sexism in science. By protesting against Wilson in his classroom, publishing editorials, and giving public lectures, the SSG employed a moralized public shaming in an attempt to excise racism and sexism from the practice of science.

What is undeniable is that Wilson's experience of public criticism helped him develop a scientific liberalism that would be associated with the specific evolutionary ideas within *Sociobiology*. And although Wilson's liberalism would be expressed as a complete political philosophy only in later decades, even in the first edition of *Sociobiology* he articulated a grand vision for the authority of evolutionary science. Wilson's conviction that a true understanding of human nature would come through the triumph of biology is perhaps most clear in the book's discussion of humans. The last chapter opens with an exhortation: "Let us now consider man in the free spirit of natural history, as though we were zoologists from another planet completing a catalog of social species on Earth."[24] This phrasing was revealing. It demonstrated Wilson's vision of science as free and unencumbered by society, with a method and purpose so universal that they extended to other planets.

It was in this context that *Sociobiology* discussed the evolutionary nature of gender roles. When Dawkins published *The Selfish Gene*, which many in the SSG saw as a direct follow-up to Wilson's work, in 1976, he included a chapter titled "Battle of the Sexes," in which he argued that males and females were locked in an evolutionary struggle over sexual fidelity.[25] Many women in the natural and social sciences regarded the writings of Wilson and Dawkins as fueling the very stereotypes they had had to battle to gain entry to the academy. The sociobiological conviction that gender was biological touched a cultural nerve at the very moment when the grassroots women's movement was developing a strident opposition to masculine authority over women's health and reproductive agency.

Genes and Gender

Well before the era of organized social protests for women's legal rights, women in North America had reckoned with the role that science and medicine played in shaping their lives. The crystallization of modern biological definitions of identity during the nineteenth century had only clarified a vocabulary for what many women already knew through experience. Since the nineteenth century, women had sought both to acquire and to dismantle the power of scientific and medical authority.[26] But because the relationship between science and women's liberation was intimately tied to their classification by race, class, and sexuality, the same knowledge that could be used to empower Anglo-American women could oppress immigrant women from Eastern Europe or Asia and subjugate Native and Black women. Beginning in the late 1960s, however, a new political energy emerged in the grassroots efforts of the women's health movement. Across the country, in a wide variety of communities, women created health clinics to offers service for women by women, including routine gynecological exams, pregnancy care, contraceptives, and abortions.[27] Women pushed back against the condescension and gatekeeping of male physicians and spoke of the empowerment they felt when they checked their own cervixes or controlled their own fertility.

It was this women's health movement that provided the political framework for the academic criticism of the portrayal of sex differences in sociobiology. This criticism was led by feminist scientists, who initially focused on the specifics of Wilson's text to highlight the lack of data in his discussion of gender roles and other human social behaviors. Over time, they also developed new humanistic methods to critique the influence of sexism in scientific practice and epistemology. In this, they drew upon some of the same historical and philosophical arguments that Gould had deployed in his criticisms of race in earlier *Natural History* columns. But the work of these scientists would have a more tangible disciplinary legacy than Gould's writings in the subsequent development of feminist science studies, for their writings explored the more general question whether, and how, science could be liberatory for women.

The most influential feminist critiques of sociobiology emerged from the Genes and Gender Collective.[28] The group got its start by forging connections between women's committees within professional science organizations, including the American Museum of Natural History and the New York Academy of Sciences. Ethel Tobach, a researcher and curator at the American Museum of Natural History, along with Betty Rosoff, an endocrinologist and professor of biology at the Stern College of Yeshiva University, organized an initial conference in January 1977. The pamphlet for the first symposium revealed the collective's concern about social confidence in "the idea that 'genes determine destiny.'" Obliquely referencing works by Wilson, Dawkins, and other sociobiologists, the pamphlet declared that "sexism is at the core" of "many recent publications" that viewed "women's social roles as 'biologically' or 'genetically' determined."[29] More than three hundred people gathered for an event that Rosoff and Tobach had planned for fifty, and many attendees spoke of the urgency and timeliness of the symposium.[30] The collective continued holding conferences and publishing volumes through the early 1990s, all devoted to issues in gender and science.

The early years of the collective provide insight into what a leftist, women-led approach to the science of sex difference looked like in this period. All the papers published in the collective's first volume were written by women trained in the natural and social sciences who had active research programs in such fields as biology, physical anthropology, and chemistry.[31] They largely framed their specific objections to sociobiology, as well as their more general concerns about hereditarian thinking in the biology of gender, as criticisms made by scientists of other scientists on scientific grounds. For instance, Anne Briscoe, a biochemist at Columbia University, argued in her chapter, "Hormones and Gender," that the "hormonal endowment in both sexes is qualitatively comparable" and that "the endocrinology of sex is not separate" in the human species, thereby countering sociobiology's faith in the discontinuous action of sex hormones in differently sexed bodies.[32] Briscoe was one of the founding members of the Association for Women in Science, which was in the process of becoming the leading national organization advocating for the professional concerns of women in the natural sciences.[33] At the time of the Genes and Gender conference, she had just finished two years as the association's president.[34]

Another contributor to the volume was Eleanor Leacock, a leading voice in the developing fields of Marxist and feminist anthropology. As a doctoral student at Columbia University in the mid-1950s, she had written her dissertation on the kinship practices of the Innu people in Labrador while taking care of two small children.[35] Throughout her scholarship, Leacock roundly criticized the prevailing intuitions that hunting and gathering societies were patrilineal and patriarchal. Her active engagement with Black activist groups shaped her work in the 1960s and 1970s, including a critique of the "culture of poverty" thesis in 1970.[36] In her chapter for *Genes and Gender*, she took Wilson to task for his superficial use of anthropological data to substantiate his arguments about the evolution of sex-specific behavior. Leacock specifically critiqued Wilson's misinterpretation of empirical studies of hunter-gatherer societies.[37] According to Leacock, the anthropological data supported a very different picture of prehistoric gender relations than Wilson provided, one in which women did "not stay in camp" but "dug roots, gathered seeds, fruits and nuts, and/or collected shellfish."[38] Leacock's critique of sociobiology highlighted the limits of Wilson's use of empirical data, even as he claimed the mantle of scientific authority.

Leacock, Briscoe, and the other authors in the first volume of *Genes and Gender* addressed the broader political implications of the sociobiological claims about biological sex difference, but they did so largely through point-by-point refutations of the relevant data behind Wilson's and other sociobiologists' scientific assertions. From the wider context of their social activism and professional lives, however, it is clear that the initial work of the Genes and Gender Collective drew upon a broad set of feminist concerns about the relationship between science and gender roles.[39] By the second *Genes and Gender* volume, the focus and methods of the collective's work had expanded. The new organizers, Marian Lowe and Ruth Hubbard, had a keen desire for the volume to be seen as something that "contained much beyond sociobiology."[40]

Lowe and Hubbard had been trained as natural scientists, in environmental chemistry and biology, respectively, but they soon emerged as distinctive voices in a new humanistic and social scientific literature that criticized Western science for its traditional gender bias. Born to a Jewish family in Austria, Hubbard fled to the United States as a teenager in 1938, in the wake of the Nazis' annexation of her native country, and earned her PhD from Radcliffe College in 1959.[41] When she later reflected on the early years of her biological work in her 1990 book *The Politics of Women's Biology*, Hubbard explained that initially she had not questioned the assumptions of her professional culture. But in the mid-1970s, as the Vietnam War and the women's movement had led her to "look closely at these assumptions," she had seen the urgent need to combat the "biological theories [that] began to be revived that threaten women's struggle for equality."[42] And Lowe was a chemist who taught regularly at Boston College, where she also directed the women's studies program.

This interest in connecting science to the broader struggles of women throughout Western history was evident in the introduction to *Genes and Gender II*, which Hubbard and Lowe coedited. In their introduction, Hubbard and Lowe employed historical context and argumentation to develop a broad critique of hereditarian explanations for gender roles. Beginning with the early-nineteenth-century view that a woman's nature was fundamentally determined by "her reproductive system," Hubbard and Lowe described the subsequent rise of evolutionary biosocial theories and, later, a science of innate sex differences. These ideas, according to Hubbard and Lowe, had been used "not only to explain but also to prescribe proper female behavior."[43]

Throughout their account, Hubbard and Lowe referred to accounts of scientific racism and sexism, arguing that both were part of a reactionary politics of biological determinism. Their points were patterned along the same argumentative lines as those advanced in Gould's *Natural History* columns on race science a few years earlier. Like Gould, Hubbard and Lowe depicted scientific sexism and racism as fundamentally conservative political enterprises. In his 1974 column "The Nonscience of Human Nature" Gould had argued that "the political function of biological determinism has been to serve the supporters of class, sex, and race distinctions. . . . In the context of Western history, this means that biological determinism has served as a tool of state and commercial power."[44] In a similar vein, Hubbard and Lowe contended that the renewed scientific interest in sex difference and race difference had "appeared alongside . . . the many challenges to the existing social structure" represented by "the women's movement" and "the demands for equality on the part of blacks and other minorities."[45] They directly cited Gould's writings on race science in *Natural History* and *Science* to substantiate these links.[46]

At one level, the work of the Genes and Gender Collective paralleled the political tactics of a radical science group such as Science for the People. As we have seen, SftP was focused primarily on disruptions to professional academic science, protesting at conferences or writing criticisms of scientists in the SftP newsletter. Similarly, much of the work of the Genes and Gender Collective happened in the spaces of professional academia. The collective held conferences, published proceedings, and created new women's forums in professional societies in the natural sciences. But all these efforts revealed a deep suspicion about the capacity for science and medicine to be tools for women's liberation even if women succeeded in diversifying the gender representation of STEM fields. Many of these feminist figures—Leacock, for instance—were more broadly critical of scientific knowledge, questioning its role as a foundational epistemology in Western societies.

A more profound critique of science can also be seen in some of Hubbard's later work, particularly as the 1970s gave way to the conservative retrenchment of the early 1980s. As the first woman to be tenured in biology at Harvard University, Hubbard was widely regarded during her career as a unique force among feminist objectors to biological theories of women's inequality.[47] Much of her writing in this period was in a similar vein to the efforts of the *Genes and Gender* volume she coedited. For instance, in her 1982 essay "Have Only Men Evolved?" she argued that the patriarchal context of Darwin's Victorian origins continued to limit the empirical strength of evolutionary theories of sex and gender.[48] She described the male-centered science of current evolutionary thinking as full of "false facts" that only served to support "the sexual and social stereotypes" about women and prevented the field from truly finding a way to "explore who we are."[49] Here Hubbard expressed a view similar to that developed by Lewontin and Gould in their analyses of racial bias in the history of biology, that is, that critiquing social bias would support the development of more accurate and empirically compelling science. And this better science, in turn, could be a means to achieving social justice.

But it was in work that was explicitly connected to the women's health movement that Hubbard put forward a more radical perspective—and a potential rejection of modern science and technology. In 1984 Hubbard coauthored a chapter in *The New Our Bodies, Ourselves* titled "New Reproductive Technologies."[50] Hubbard published the piece along with Wendy Sanford, a founding member of Our Bodies Ourselves, a feminist health collective that had begun in 1969. The first edition of *Our Bodies, Ourselves* had been published by the collective in 1970.[51] Over the course of its many editions the enormously popular *book* had a far-reaching impact on health care policies and research on women's health in the United States.[52] Hubbard's chapter in the 1984 edition announced her public opposition to research into reproductive technologies, particularly in vitro fertilization (IVF), to the large *Our Bodies, Ourselves* audience. As the gender studies scholar and historian of science Jenna Tonn has argued, Hubbard objected to IVF on the grounds that the technology would reify women's biological destiny to be mothers and, moreover, hand the medical care of women's bodies over to a male-dominated science.[53] Hubbard called for a halt to funding for research into IVF, arguing that it would be better to concentrate limited resources on means that would improve the health and lives of all mothers and children rather than on developing a reproductive technology that would aid only a chosen few. Thus, Hubbard's rejection of IVF emerged out of a context that prioritized the embodied experiences of women's health, especially with regard to their reproductive lives, and that harbored a great skepticism about the capacity for science to be part of women's political and social liberation.

The tension between reform and a more radical rejection of modern science would continue as the field of feminist science studies expanded in subsequent decades. At issue was whether the assimilation of women into the practice of science—that is, having more women doing research in biology, chemistry, physical anthropology, and psychology—would be enough to lift science out of its history of sexual and gender oppression. More radical thinkers argued that achieving gender equality would require cultivating a greater skepticism toward the epistemology of modern science. But before turning to these developments, let us consider how Gould criticized Wilson by deploying the concept of biological determinism and how Wilson, in turn, developed scientific liberalism as a defense against this critique.

Biological Determinism

The first few years of Gould and Wilson's relationship at Harvard were cordial. Their early exchanges were similar to Gould's interactions with his other colleagues in the Harvard biology department during this period.[54] Yet, signs of trouble were brewing. By the time Wilson's book was published in 1975, Gould had already written several pieces for *Natural History* on racism and biological determinism. In the spring of 1976 he decided to address the sociobiology controversy directly in a column titled "Biological Potential vs. Biological Determinism." In the piece, Gould asserted that he had taken up the subject because of the "intense discussion" and "chorus of praise and publicity" aroused by the publication of *Sociobiology*.[55] He warned his readers to be wary of the new theory since biological determinism had "always been used to defend existing social arrangements as biologically inevitable." He further argued that while individual scientists might not see themselves as architects of oppression, the use of biological determinism to justify racism and sexism was "quite out of the control of individual scientists who propose deterministic theories for a host of reasons, often benevolent."[56] The only possible way to eradicate what Gould depicted as the misuse of science was to recognize the influence of social biases on science.

After this prelude on the dangers of biological determinism, Gould's column largely focused on specific critiques of sociobiology. First, he argued that the sociobiological approach to evolution wrongly atomized organisms into a series of traits, each matched with a single gene subject to natural selection. He would go on in 1979 to expand this argument in greater detail and to greater infamy in "The Spandrels of San Marco." But in 1976 Gould had not fully worked out the rhetorical flourish that would characterize the latter paper.[57] Gould's second critique focused on Wilson's lack of evidence for the genetic underpinnings of human social behavior. In Gould's account, Wilson had relied on three assumptions about human behavior rather than scientific evidence to develop his account of human sociality: that human behaviors were universal, continuous, and adaptive. In saying that Wilson assumed human behaviors were universal, Gould meant that *Sociobiology* inferred that social behaviors were genetic because they were found across all human societies. To refute this point, Gould simply stated, "I can only say that my own experience does not correspond with Wilson's."[58] Gould's reference to "continuity" highlighted *Sociobiology*'s reliance on analogies between human and animal societies, assuming, in other words, that evolution was continuous. Gould acknowledged that the theory of reciprocal altruism was a well-evidenced explanation for other-oriented behaviors in many animals, but he disagreed with the sociobiological perspective that "humans perform altruistic acts and these are likely to have a similarly direct genetic basis."[59] Finally, Gould contended that Wilson's characterization of human sociality rested upon an assumption that individual human traits were adaptive, an assumption that Gould believed existed in a "context of no evidence."[60] Each of Gould's criticisms stressed Wilson's lack of empirical proof for his claims about the evolution of human social behavior.

Gould wanted to add his voice to criticisms of Wilson's book, hence his column. But he also wanted to maintain a cordial professional relationship with his colleague, so he sent Wilson an early draft of the column in March 1976. He sent an accompanying letter requesting a response: "I debated intensely with myself before writing [the May column] at all. . . . Nonetheless, I realized one day that I am the only popularist writing a regular feature on evolution—how could I ignore the most widely discussed event . . . in evolutionary biology during my brief career[?]"[61] Gould's letter to Wilson balanced flattery and self-aggrandizement. In this contentious moment, Gould emphasized what he believed was his ethical obligation as a public intellectual over his identity as a professor of biology and an academic colleague.

Wilson utterly rejected Gould's framing of the matter. In his response to Gould, Wilson declared that he was "appalled to receive the article." He found Gould's decision to air the debate in the pages of a popular magazine professionally unsound and morally questionable. Wilson informed Gould that "considerable harm is being done to evolutionary biology here because of the highly personal nature" of the "bitter struggle over the political issue of sociobiology."[62] Wilson vehemently criticized Gould for "continuing a partisan attack, essentially identical in its arguments to that used by Science for the People, in an important forum where I will have no chance to reply."[63]

Gould attempted to defuse the situation, telling Wilson, "I can only say that I feel no personal animosity towards you whatsoever, and that I would welcome a renewed personal contact with you." He also assured his colleague that he had meant for Wilson to have the opportunity to write a reply for *Natural History*.[64] Wilson took Gould up on the offer to draft a reply, but the letter he sent to the editor ignored Gould's scientific critiques. Instead, Wilson informed the readers of *Natural History* of Gould's activist and political orientation: "Readers not already familiar with the story will find that Mr. Gould is a member of Science for the People, a radical activist group which has been conducting an intense campaign to discredit any notion of a genetic foundation of human nature." Wilson defended his interest in sociobiological research, writing, "Readers . . . are asked to remember that the many biologists and social scientists interested in the genetic basis of human nature are just as humanitarian as the tiny minority who consider the subject taboo."[65] According to Wilson, biologists were sufficiently responsible to regulate the ethics of their own proceedings. Wilson depicted Gould as part of a vocal but small group casting doubt on what Wilson characterized as a relatively established scientific consensus.

Gould, outraged, returned the favor by writing a response that began by stating that Wilson "has chosen totally to ignore the biological content of my article." He objected, as well, to having his views clumped in with those of Science for the People, which he described as "a large collective group, which I neither lead nor serve as primus inter pares."[66] Gould, in other words, turned Wilson's critiques back on him. Gould concluded his response by expressing his distaste for both sides of the affair.[67] In the end, Gould's editor, Alan Ternes, elected to publish only Gould's column in *Natural History*.

In the decades since, sociologists, historians, and the two men's colleagues have offered various theories for the depth of the divide that opened between Wilson and his critics. Gould and Wilson maintained different approaches to the role of adaptive explanations for evolutionary phenomena—a difference that would only sharpen after 1979. Moreover, Wilson did not approach progressive political concerns with the respect and deference that Lewontin, Beckwith, Gould, and like-minded biologists exhibited. Undoubtedly, this combination of technical and political differences sparked the sociobiology controversy immediately after the book's publication and kept it going for several decades. But private correspondence between Gould and Wilson demonstrates that their conflict turned on a much more specific issue, namely, scientists' ethical duty to the wider public. Gould contended that scientists should be self-reflective and forthright about the potential limitations and biases of the scientific process. He was convinced that this approach was the only safeguard preventing scientists from contributing to a history of scientifically justified social oppression and violence. Wilson was persuaded of just the opposite; he contended that scientists ought to convey the truth about the natural world to the wider public without weighing science down with any political or social considerations.

Relations between Gould and Wilson continued to be strained for the next few years. They deteriorated when the conflict escalated beyond the polite chambers of meetings and teach-ins. At the annual meeting of the American Association for the Advancement of Science in February 1978, during a session on sociobiology, a group of protestors from the Committee Against Racism doused Wilson with a bucket of water.[68] Although Gould publicly deplored the physical nature of the confrontation, Wilson wrote a chiding letter questioning whether Gould was "prepared to be as stern with Science for the People, even after you disassociated yourself with that group."[69] Despite Wilson's skepticism, Gould was upset enough by the events to terminate his public criticism of Wilson. Subsequently, he would find subtler ways to continue his disagreements with the explanatory premises of sociobiology. One way was through *The Mismeasure of Man*.

From Sociobiology to *The Mismeasure of Man*

During his disagreements with Wilson, Gould was at work, along with his colleague Garland Allen, on a book-length treatment of biological determinism in the history of science. They intended the book to serve as the ultimate reproof to *So-*

ciobiology. Allen was an outspoken leftist whom Gould had met in the context of transformations to the discipline of the history of science at Harvard and the creation of the history of biology as an independent subdiscipline. Gould and Allen became good friends; indeed, Gould wrote many of his columns for *Natural History* in Allen's house on the shore at Woods Hole, Massachusetts.

In the spring of 1976, Gould and Allen sent a proposal for a project they were calling *Biological Determinism* to Gould's editor, Edwin Barber. They proposed a social and political history of biology organized chronologically to describe instances of biological determinism in the Anglophone world from the mid-nineteenth century through the sociobiology controversy. The two men would be responsible for different sections, with Allen focusing on eugenics and Gould concentrating on nineteenth-century craniometry. In the proposal, Gould and Allen claimed that "biological determinism" was to blame for "the evils of slavery, the prevention of women's suffrage, the exploitation of the poor . . ." and the "policies of eugenics." According to Allen and Gould, the ideology of biological determinism was directly responsible for the historical and existing inequalities of race, gender, and class in American history. By setting contemporary research, including sociobiology, into a longer history of biological difference, Gould and Allen implicitly ridiculed the contemporary science for its biases and its hubris: "Whether the biologically determining factors are deemed to be genes, hormones, or the shape of the skull, they are viewed as innate from the time of conception."[70] According to their argument, these current theories held no more credence than the passé theories from earlier times.

The proposal was successful, and Gould and Allen secured a contract with Norton. A series of letters between the two during 1976 show them exchanging reports on their progress amid bits of news about the house at Woods Hole. But differences soon arose. "I'm sorry we don't agree about the whole sociobiology affair," Allen wrote to Gould in February 1977. Although Allen reiterated his feelings of friendship for Gould throughout, the letter was firm about his beliefs, expressing his concern that Gould was "more outraged by the personal indignities cast upon Wilson" than about "the real potential danger of sociobiological ideas themselves." With Gould increasingly hesitant to castigate Wilson in public, it seemed to Allen that Gould was missing the central point of what they had criticized about sociobiology. As Allen wrote, "I feel you don't take the political clout of academic ideas seriously enough. It may be true that Wilson is a nice guy who would never hurt anyone directly; he may not be personally racist or even sexist; but he has put himself on the front line as the architect of a major theory of human limitations."[71]

Despite these differences, work on the book continued. But as the months passed, the correspondence between the two friends revealed an unraveling of the book project. By early 1979 the idea of a collaborative book project had dissolved entirely.[72] Allen and Gould, despite seemingly good-faith efforts on both sides, could not agree on the extent to which the project should be explicitly political. The two men took the portions of the project that they had been working on and went their separate ways. Allen's portions of the proposal had a different life in his writings in the history of science, including his discipline-changing work on American eugenics,[73] while Gould went on to write what would become his most famous ethical treatise on science, *The Mismeasure of Man.*

Most readers today encounter *The Mismeasure of Man* as a takedown of so-called race science, especially ideas about intelligence. The published volume nevertheless retains traces of its roots as a riposte to Wilson's *Sociobiology*. The introduction contains an oblique reference to the "current resurgence of biological determinism" but explains that the "ephemeral" nature of such claims makes them more appropriate for discussion in magazines or newspapers than in a work of history.[74] Otherwise, Gould confined his remarks on the "brouhaha over sociobiology" to the conclusion of the book, in which he returned to the argument he had laid out in his 1976 column.[75] This time, he directly labeled "the search among specific behaviors for the genetic basis of human nature" as a form of biological determinism. Gould went on to argue that while we ought to view our brains in their entirety as products of evolution, we should not describe individual psychological traits as results of the selective process.[76] But even with these passages, *The Mismeasure of Man* was not generally regarded as part of the sociobiology debate. Norton's publicity for the book made no mention of sociobiology, instead advertising it as a natural follow-up to Gould's *Natural History* anthologies, nor would most audiences in subsequent decades connect Gould's book to his altercations with Wilson.[77]

It seems that most readers focused on text, not subtext. The bulk of *The Mismeasure of Man* analyzed episodes in American history in which scientists used quantitative and psychological data to claim innate differences in intelligence between racial groups. Gould argued that it was this process of quantification, combined with the "abstraction of intelligence as a single entity," that had given scientists the confidence to "rank people in a single scale of worthiness." The first half of the book focused on nineteenth-century craniometry, while the latter half examined twentieth-century intelligence testing, particularly the development of the intelligence quotient (IQ) test. Gould highlighted moments in these episodes in which he believed racist assumptions had directly influenced scientific practices. His point, however, was not that racism turned scientists into mustache-twirling villains. He insisted that he did not mean to "contrast evil determinists" with those "who approach data with an open mind and therefore see the truth." Rather, he criticized "the myth that science itself is an objective enterprise."[78] For the rest of his career, Gould would continue to define ethical science in these terms. He believed that scientists, through conscious self-reflection, could avoid the biases that inevitably crept into the work of science.

Although nineteenth-century accounts of race and intelligence might appear to have a tenuous connection to Gould's work in paleontology, there was in fact a substantive link between the two. After all, the relationship between racial variation and the trajectory of evolutionary history had been a central preoccupation in paleontology since before the twentieth century.[79] Gould united these two themes in his criticism of evolutionary narratives of racial progress. His dislike of this view of the evolutionary process grew out of his research on the relationship between individual development and evolutionary change, the subject of his first technical monograph, *Ontogeny and Phylogeny*. One of Gould's critical contentions in that book was that pre-Darwinian embryological studies fell into two strains. On one side, a series of early-nineteenth-century thinkers had argued that the stages of an individual's growth and development (ontogeny) repeated "the *adult* forms of animals lower down the scale of organization" (phylogeny). This "biogenetic law" focused on a specific and fixed relationship between individual development and the evolutionary process. Gould argued that when this theory was "reread in the light of evolution" it set the stage for scientific racism.[80]

In both *Ontogeny and Phylogeny* and *The Mismeasure of Man*, Gould cited the work of the nineteenth-century Italian physician and criminologist Cesare Lombroso in order to argue this point. In Lombroso's work, wrote Gould, criminals were marked by their "apish or primitive human features," including "relatively long arms, prehensile foot with mobile big toe, low and narrow forehead, large ears, thick skull, large and prognathous jaw, copious hair on the male chest, browner skin."[81] According to Lombroso, the link between savages, criminals, and children was not merely metaphorical; he ranked their place on the evolutionary ladder by their social maturity. Importantly, in *The Mismeasure of Man* Gould stressed that Lombroso's scientific racism stemmed from the "specific *evolutionary* theory" he used, not "just a vague proclamation that crime is hereditary."[82] In other words, Gould argued, a progressive view of evolution had produced some of the most pernicious forms of scientific racism during the nineteenth century.

In *The Mismeasure of Man* Gould concentrated on the relationship between social prejudice and scientific racism. But he also had in mind a second evil, which was the harm to evolutionary science itself caused by this type of progressive thinking. The nineteenth-century belief that ontogeny recapitulates phylogeny not only encouraged racism, Gould contended, but also caused scientists to miss the myriad possibilities for understanding the relationship between ontogeny and phylogeny. Part of his effort in *Ontogeny and Phylogeny*, therefore, was to revisit another strain of pre-Darwinian thinking, one that had looked not to a specific, fixed relationship between individual development and evolutionary results but instead to the myriad relationships between the two processes. Gould argued that it was possible to focus on process and therefore revive the scientific study of the relationship between ontogeny and phylogeny.[83] In his arguments, he presented racism as not just a social evil but a social bias that had been detrimental to the advancement of science.

What, then, was the path forward? For Gould, one significant possibility was historical reflection. He explained that he had turned to history because it illustrated "some generalities about science" that would "surprise no historian but prove interesting to many scientists." Critically, he asserted that "theory must play a role in guiding observation, and theory will not fall on the basis of data accumulated in its own light."[84] In his view, contemporary scientists could use a careful historical study of older theories to identify bias and move their own research forward. Gould's comments on the limitations of progressive thinking were an early example of his belief that science was theory-laden, that is, that the empirical facts of science were made sensible by preexisting conceptual frameworks, a perspective informed by the Harvard philosopher of science Thomas Kuhn's notions of scientific paradigms.[85] In keeping with Kuhn's view, Gould rejected the more traditional view that science advanced through the "simple accumulation" of data under the universal aegis of the scientific method. In the case of progressive thinking in evolution, Gould argued that this mistaken assumption hindered evolutionary biologists from considering the rich area of research possible on ontogeny and phylogeny.

Gould's reliance on the theory-ladenness of science shaped his analysis of the history of racism and science throughout *The Mismeasure of Man*. This was true, for instance, in his assessment of Samuel George Morton's work. Morton's *Crania Americana*, published in 1839, drew upon the assumption that the human races were separately created species.[86] Equating brain size with intelligence, Morton used his extensive skull collection to argue that "Caucasians" had larger skull capacities and were therefore inherently more intelligent than "Negroes." In what would become the most infamous portion of *The Mismeasure of Man*, Gould argued that Morton's racist biases had influenced his measurement and analysis of the skulls. Gould did not actually remeasure Morton's skulls; instead, he worked with the data in *Crania Americana* in an attempt to discern whether Morton's measurements reflected a bias toward higher intelligence in the Caucasian race. As he informed his readers, he spent several weeks during the summer of 1977 reanalyzing by hand on legal pad paper the raw data of the skull measurements that appeared in Morton's volumes.[87]

Gould could find "no evidence of conscious fraud"; in fact, he believed that Morton would not have published the raw data had he been a "a conscious fudger." Instead, Gould argued that the unconscious nature of Morton's bias was entirely the point. As he explained, "The prevalence of unconscious finagling, on the other hand, suggests a general conclusion about the social context of science. . . . For if scientists can be honestly self-deluded to Morton's extent then prior prejudice may be found anywhere, even in the basics of measuring bones and toting sums."[88] Gould contended that Morton's desire to uphold the antebellum racial hierarchy had dictated his empirical conclusions.

With *The Mismeasure of Man*, Gould developed an approach to scientific racism that was meant to address social inequality while also producing more reliable science. Gould used historical reflection on racial bias not as a way of discrediting science but as a tactic that could shift scientific paradigms away from racism. Although the liberal defense of science would come to see this emphasis on racism as an attack on the authority of science, a close reading of *The Mismeasure of Man* suggests that this was not Gould's intention. A humanistic approach to science, in Gould's hands, was intended both to direct evolutionary thinking toward greater truth and to champion social equality.

Feminist Science Studies

Gould was hardly the only scholar to suggest that engaging with the science of biological difference was an act of justice. Throughout the academy, feminist approaches to science, whether developed by scholars trained in the sciences or the humanities, were by far the most influential in tying together political activism and scientific epistemology. These scholars developed a complex set of arguments about bias in the practice of science and challenged implicit Western hierarchies that enshrined the natural sciences above other ways of knowing. Although often caricatured as such, these feminist perspectives cannot be characterized as simple relativism, nor did they focus exclusively on masculinist bias in science. Rather, feminists developed a more fundamental set of epistemological and ontological questions about knowledge that would have far-reaching consequences for the development of critical theories throughout the humanities.

In the decades following the publication of *The Mismeasure of Man*, feminist historians of science cited Gould's work in their historical critiques of modern science. Cynthia Russett's 1989 book *Sexual Science: The Victorian Construction of Womanhood*, for example, cited both *Ontogeny and Phylogeny* and *The Mismeasure of Man* in its analysis of notions of evolutionary progress in late-nineteenth-century debates over women's higher education.[89] *Sexual Science* attracted wide attention for the precision of Russett's analysis of Victorian science and for the clarity of her arguments regarding the sexual stereotypes that animated Darwin's theory of evolution.[90] A few years later, Londa Schiebinger, in her 1993 volume *Nature's Body: Gender in the Making of Modern Science*, referred to Gould's *Natural History* column in an examination of gendered language in Linnaean taxonomy.[91] By demonstrating how the gendering of scientific knowledge was at the core of its history, Schiebinger's work in *Nature's Body* was instrumental in establishing gender and science as an integral perspective in the field of the history of science. The shared sympathies between Gould's public writings and these works, however, should not be interpreted as evidence that Gould was at the center of or even a significant participant in the feminist history of science or feminist science studies. Even in the case of sociobiology, Gould's technical and public criticisms concentrated on racial hierarchies, not social gender roles or sexuality.

In the writings of Russett, Schiebinger, and others, history of science became a kind of critical feminist practice. Historical analysis opened up the process by which contemporary science had emerged out of racial and gendered assumptions and made possible the critique of biases woven into supposedly value-neutral science. Importantly, history had also been deployed as a mobilizing tool in the wider women's movement. It had been used, for instance, by the Women's Action Alliance, the feminist organization founded by Brenda Feigen Fasteau, Dorothy Pitman-Hughes, and Gloria Steinem in 1971. From its inception, the WAA was a national leader in feminist organizing, connecting networks of grassroots organizations with strategies, information, and public narratives.[92] In its early years the organization carried out a number of projects that made history an integral part of feminist practice. One such project in the summer of 1979 was the Institute of Women's History, held at Sarah Lawrence College in cooperation with the

Smithsonian Institution.[93] The intention of the event was to make women's history an integral part of the consciousness of women's organizations and to raise national awareness of women's history.

This conviction that historical literacy could help drive social change also influenced Russett and Schiebinger's work. Although their scholarship was created within and for the academy, textual traces reveal an orientation toward using this historical work to engender political change. In the introduction to *Sexual Science* Russett commented on the relevance of Victorian history in a present in which "sex and scientific continue their uneasy relationship" in fields such as "hormonal research . . . and sociobiology."[94] Drawing a connection between the nineteenth century and the present, she argued that the twentieth century had not succeeded in building an equitable world and that "the dismal history of sexual science" suggested that "women and other marginal groups" were right to "continue to view the scientific pursuit of group and individual differences with suspicion."[95] In a similar fashion, in *Nature's Body* Schiebinger leveraged the history of eighteenth-century European natural history in terms of the ongoing issues of the day, arguing that "ours is a culture obsessed with difference." To understand this obsession, she argued, "we must unearth its origins and its history."[96]

We can further trace a distinctly feminist approach to science by comparing special issues in two professional feminist journals, each dedicated to feminism and science, published approximately a decade apart. The first was the October 1978 issue of *Signs: A Journal of Women and Culture.* Founded in 1975, *Signs* quickly emerged as a flagship publication in feminist studies. The special issue titled "Women, Science, and Society" included the first work by several authors that looked at science specifically through the lens of academic feminism.[97] The introduction to the issue made clear that the question of science in relation to women was relatively new, stating that the issue "takes up a subject—science—that is too often neglected or distorted by the new scholarship on women."[98] Some members of the Genes and Gender Collective, including Marian Lowe, had pieces in it, while others were written by women who would come to be among the most significant in the field, including the historian of science Sally Gregory Kohlstedt and the feminist theorist Donna Haraway.[99]

The argumentative methods in the special issue of *Signs* overlapped with those in the first two *Genes and Gender* volumes. In fact, Lowe's piece in the *Signs* issue was partially adapted for the second *Genes and Gender* volume. In it, she criticized sociobiology on both its scientific merit and its potential political impact, reprising a prominent perspective from these early feminist critiques of science: that sociobiology and other forms of biological determinism were bad science according to the terms of science.[100] However, the *Signs* special issue also contained pieces that set the foundation for comprehensive historical and philosophical critiques of sciences. Kohlstedt contributed a piece titled "In from the Periphery: American Women in Science, 1830–1880," in which she analyzed women scientists' efforts to participate in a rapidly professionalizing scientific community. She concluded that earlier generations of women "working on the periphery provided the foundation" that allowed for later generations to "access advance training and scientific associations."[101] Haraway's contribution explored the relationship between political and physiological theory during the eighteenth century, arguing that feminists needed to take bolder steps in questioning the foundations of Western science. She averred that even though feminists had been "less mesmerized than many" by science's claim to "value-free truth," they had misunderstood "the status and function of natural knowledge." In this, she claimed that feminists had been complicit in granting "science the role of a fetish, an object human beings make only to forget their role in creating it."[102] According to Haraway, only through a thorough feminist engagement with the heart of science could science be reimagined for justice.

The second special issue, "Feminism and Science" in *Hypatia*, was published in two parts, in the fall of 1987 and the spring of 1988. The explosion of literature in the intervening period meant that the issue editor, Nancy Tuana, could refer in her introduction to "the feminist critiques of science developed over the last decade," in which feminists had "uncovered the complex interaction of sexist, racist, and classist biases grounding theories of human nature, and in doing so have seen the ways in which such biases permeate the entire structure of science."[103] Even the difference in the titles revealed a shift in the intervening decade. The title of the *Signs* issue focused on an object of study—that is, the question of women and science—whereas the *Hypatia* title invoked a theoretical point of view. In the intervening period, feminists had developed an influential and critical approach to knowledge that made them willing to question the entire epistemological framework of Western science. The contributors to the *Hypatia* issue asked the most essential and profound questions about scientific knowledge, including whether it could be understood as objective, reliable, or trustworthy.

Two thinkers in the *Hypatia* issue brought forward these philosophical questions with great clarity. These were Helen Longino, who contributed "Can There be a Feminist Science?," and Evelyn Fox Keller, author of "The Gender/Science System." Longino, who held degrees in literature and philosophy, was active in antiwar and women's liberation activism through the 1960s and 1970s.[104] In

"Can There Be a Feminist Science?" Longino made two arguments that she would expand on in her 1990 book *Science as Social Knowledge: Values and Objectivity in Scientific Inquiry*. Longino criticized the assumption that a feminist science would be a feminine science characterized by specific qualities, such as "complexity, interaction and wholism," associated with and meant to valorize the "female sensibility or cognitive temperament."[105] Instead, she contended that it was necessary to approach science as "practice rather than content . . . hence, not [as] feminist science, but doing science as a feminist."[106] Longino claimed that feminist interventions to masculinist bias should not be seen as an outside force in science but rather as part of a new "business as usual" in which the scientific community reflected on and recognized the "deep metaphysical and normative commitments" expressed by any science in the "culture in which it flourishes."[107] For Longino, feminist engagements with science would ideally be specific, local, and targeted critiques that would produce concrete and practical improvements to the research practices of science.

Keller, who had been trained as a physicist, became one of the leading voices of feminist approaches to science after teaching her first women's studies course in 1974. In her *Hypatia* article, she reflected on themes that she had also explored in her 1983 book *A Feeling for the Organism: The Life and Work of Barbara McClintock*. In that book Keller had examined the influence of McClintock's gender on her microbiological practices; Barbara McClintock had won a Nobel prize in 1983 for research on transposable genes that she had begun thirty years earlier. In her *Hypatia* article, Keller revisited the influence of gender identity on science and the question whether Barbara McClintock was an exemplar of feminist science. Echoing Longino, Keller argued against valorizing feminine science defined by specific feminist qualities, but she also wished to acknowledge that gender ideology was inescapable. In the case of McClintock, Keller believed that even though McClintock was not a feminist and in fact believed that science was a realm in which gender was irrelevant, the pervasiveness of gender ideology meant that it still influenced McClintock's work.

These *Hypatia* articles constituted a full-scale reassessment of how science should be conceptualized. These and related works had far-reaching effects in the academy, both on scholars working directly at the intersection of feminist theory and STEM fields and on broadly held assumptions about the relationship between feminism and science. Importantly, feminist science studies developed in the early years of the modern field of women's studies and feminist theory itself. As women's studies became institutionalized, with the first department at San Diego State University in 1970 and the first doctoral program at Indiana University in 1980, feminist science studies not only influenced humanistic and social scientific theories of gender; it also influenced academic feminism overall. This way of thinking about science—with the freedom to question science's cultural neutrality, to doubt its universality and objectivity—contributed to some of the most important developments in feminist theory beyond the specifics of scientific knowledge.

An especially prominent example of these developments was Judith Butler's discussion of the body in their 1990 book *Gender Trouble: Feminism and the Subversion of Identity*. Butler's background was in literary studies, and the book established them as one of the leading figures of third-wave feminism and queer theory. The crux of Butler's argument was to break down the seeming naturalness of the relationship between sex, gender, and sexuality—that bodies with certain anatomy have specific cultural roles and particular sexual desires; that is, that men are men because they have penises and testes, which in turn means that they will act aggressively and desire women sexually. Butler argued against the naturalism of these connections, contending instead that we culturally understand men to be heterosexual and masculine because of repeated patterns of social behavior that iteratively cite earlier forms of the same behavior. This iterative social citational practice is what renders gender comprehensible, but it does not emerge from, and is not reducible to, the natural body. Butler referred to this practice as "gender performance."

Most of Butler's approach to the question of gender and how it relates to the body was disconnected from the specifics of the feminist science studies approach to the female body or to the science of sex difference. Much of their analysis, for instance, focused on a critique of the textual logics of French feminist psychoanalytic theory. But there was a sense of freedom in Butler's willingness to set aside the materiality of the gendered body that can be at least partially attributed to feminist science studies; notably, Haraway and Keller were among the colleagues they thanked in the preface to *Gender Trouble*.[108] In the first chapter, "Subjects of Sex/Gender/Desire," for example, Butler marshaled evidence against the assumption that sex and gender were natural to consider whether "'identity' [is] a normative ideal rather than a descriptive feature of experience."[109] Having questioned the biological justifications typically used to determine sex, they asked, "How is a feminist critic to assess the scientific discourses which purport to establish such 'facts' for us?" Butler cited several feminist science studies texts, including *Genes and Gender* volumes 1 and 2, the *Hypatia* special issue, and others, including Sandra Harding, Haraway, and Keller, to justify their use of quotes to denigrate the authority of scientific "facts." Referencing this body of literature,

Butler contended that "a great deal of feminist research has been conducted within the fields of biology and the history of science that assess the political interests inherent in the various discriminatory procedures that establish the scientific basis for sex."[110] Confident in the feminist critique of science, Butler was free to dismiss the authority of natural science over the body.

Certainly, Butler's work was not solely the product of feminist engagements with the science of sex and gender. But the sense of liberty in Butler's work, as well as their reliance on feminist accounts of sexual biology, demonstrated the maturity of the field and the power of its epistemological tools. It also promoted a more pessimistic view of the science of biological difference than Gould had offered in *The Mismeasure of Man*. Gould's account of nineteenth-century craniometry had held the promise that scientific racism could be overcome, and science made more reliable, through goodwill and historical reflection. But by the end of the century, the writings of Butler and other feminist and queer theorists had helped solidify a view within feminist studies that any claim about biological sex difference was an impediment to gender justice. In turn, this had reinforced a narrative from scientific liberals that feminist studies itself was antiscience and antibiology.

The Limits of a Critique of Biological Determinism

In 1979 the self-described "black, lesbian, mother, warrior, poet" Audre Lorde was asked to make a comment at a panel at a New York University Institute for the Humanities conference celebrating the thirtieth anniversary of Simone de Beauvoir's 1949 book *The Second Sex*.[111] The influence of de Beauvoir's existential philosophy on twentieth-century feminist thought can hardly be overestimated. *The Second Sex* argued that women's liberation rested upon freedom from "reproductive slavery," implying that it was by escaping biological essentialism that women would be released from the gendered imperatives of motherhood and domesticity.[112] In commemorating *The Second Sex*, the NYU conference of more than eight hundred women captured the political energy of the moment within academic feminism.[113] By rejecting the patriarchal authority of biology as destiny, they would fight against the restrictions that prevented women from taking on equitable leadership in society, both in the United States and around the world.

But Lorde's comment on the occasion was far from laudatory. She criticized the conference for claiming to represent feminism, despite the almost complete lack of representation for lesbian and/or Third World women consciousness. She named the silences of academic lives that depended on the labor of poor women and women of color to clean houses and care for children so that feminists could attend conferences on theory. And she spoke out against the exclusions of a white feminism that did not reckon with race, sexuality, or class. Her perspective emerged from her work within and alongside the Combahee River Collective, her transnational activism against white supremacy and homophobia, and the spiritual energy of her poetic journey.[114] Lorde's lecture, "The Master's Tools Will Never Dismantle the Master's House," was subsequently published in her 1984 essay collection *Sister Outsider*.[115] The injunction in Lorde's title of what would become the most well known essay in that influential collection struck to the heart of a Black feminist imperative, that something more than mere liberal reform was needed for Black women to survive and flourish in the world.

Lorde's 1979 critique helps to clarify the strands of academic and political consciousness that fed into the sociobiology controversy, as well as those that came out of it. In particular, it helps to unpack the vision of political change that animated the SSG's and the Genes and Gender Collective's criticisms of sociobiology. These academics rejected the liberal assertion that science was only ever misused by prejudiced social agendas but ultimately remained objective and free from ideology. Instead, they embraced an understanding of science as a resolutely social process therefore subject to the frailties of those who practiced it. Characteristic of this leftist orientation was Gould's depiction of Morton in *The Mismeasure of Man* as not an intentional villain but rather a faithful empiricist whose "a priori conviction about racial ranking" had nevertheless influenced his conclusion that Caucasians had the largest cranial measurements. Reckoning with the influence of (potentially unconscious) bias was the task that Gould, others in the SSG, and the Genes and Gender Collective set themselves. In their nonviolent direct-action confrontations with the racism and sexism of sociobiology, they echoed the political sensibility of an earlier era of the civil rights movement. Theirs was a moralized calling to the scientific community to return to being their collective better selves.

But running through the work of these groups was a fundamental tension: was modern science a "master's tool" that should be fully dismantled, or could it be a powerful agent for social justice if it were freed from racism and sexism? As we have seen, the answers to these questions varied among the participants of these groups and also changed over time. Hubbard, for instance, developed a more strident rejection of scientific innovation in the context of new reproductive technologies. Gould, by contrast, became less epistemologically radical as his career progressed, declaring in 2002 that he was a "rather old-fashioned 'realist'" when it came to a confidence that science provided a reasonably faithful accounting of the "factual world."[116] Across their work, radical scientists struggled to find a balance between the science they loved and the justice they craved. But as the foundations laid by these activist movements were developed by feminist science studies, an increasingly constructivist approach to the epistemology of science emerged. And this, in turn, laid the groundwork for the debates over scientific realism in the science wars of the 1990s. More than any other field, feminist studies institutionalized radical science activism in the academy.

Lorde's essay helps us to recognize the tension between a reformist approach and a more radical dismantling of the epistemology of science. But her critique also helps us to revisit the early moments of these movements with openness and humility as we consider what silences and gaps existed in the radical science understanding of racism and sexism in science. We can reflect on the limits of biological determinism as a political framework for understanding the historical violence of scientific knowledge and institutions.

Recall that in the publications of the SSG and the Genes and Gender Collective, as well as in *The Mismeasure of Man*, biological determinism was described as a conservative political agenda. The SSG letter to the *New York Review of Books* argued that "determinist theories" survived because they "tend to provide a genetic justification of the *status quo* and of existing privileges for certain groups according to class, race, or sex."[117] The second volume of the Genes and Gender Collective also depicted determinist biological theories as the purview of reactionary agendas that rejected "the demands for equality on the part of Blacks and other minorities."[118] And Gould urged in the 1996 edition of *The Mismeasure of Man* that we "grasp the potency of biological determinism as a social weapon—for 'others' will be thereby demeaned, and their lower socioeconomic status validated."[119] Across these texts, thinkers shared an understanding that biological determinism used the authority of objective science to enforce the status quo of social hierarchies. In an era when Jensen and Shockley used biology to justify racial segregation, this was a powerful claim.

But what this analysis misses is how central science has been, and is, to progressive politics, perhaps even more so when we take seriously the emergence of biopolitical nation-states in the modern world.[120] The term *biopolitics* comes from Michel Foucault's *History of Sexuality* and his originally unpublished lecture series "Society Must Be Defended," in which he argued that in modern governance individuals do not have a direct relationship to the state but are instead only ever signatories of a population that the state manages.[121] In the biopolitics of the nation, scientific knowledge is not only used to keep things as they are; it is also deployed to optimize and improve. This is most vividly true in the connection between the scientific management of reproduction and the quest to control the racial character of the nation. For more than a century, experts and elites in the United States embraced demography and genetics as tools to address poverty, regulate immigration, and better public health. New innovations in statistics and hereditary analysis allowed public officials to visualize changes in the racial composition of the United States owing to changes in birth and death rates and fluctuations in immigration patterns that were the result of global decolonization during the Cold War. And it was industrial pharmaceuticals that gave American elites the tools they sought to control reproduction at borders, on reservations, and in urban slums.

To put into words this expansive confidence in science to engineer a better future for the nation through the better breeding of its population, we might describe the relationship between progressive politics and scientific innovation as eugenic. Eugenics pervaded the efforts of twentieth-century feminists like Margaret Sanger to guarantee voluntary motherhood and improve the American race.[122] It undergirded W. E. B. Du Bois's belief that it was the "talented tenth" of Black elite that would uplift the African American community.[123] It shaped Theodore Roosevelt's conviction that environmental conservation was the means to ensuring the health of Anglo-Americans.[124] And it supported the rejection of disability during the German measles outbreaks of the 1960s, which in turn helped turn the political tide toward the legalization of abortion.[125] Historians and literary scholars alike have shown how intrinsic eugenic thinking has been to aspects of feminist and progressive politics from the early twentieth century to the present.[126]

But in no small part because of the 1970s critique of biological determinism, which identified eugenics with reactionary politics, the academic Left struggles to confront histories of scientific violence in progressive movements. As recently as 2019, historians of American eugenics rejected an analysis that tied eugenics to the historical effort to legalize abortion in the United States.[127] Their refusal was understandable given that the analysis came from the conservative opinion of Supreme Court Justice Clarence Thomas.[128] I write now after the 2022 *Dobbs v. Jackson Women's Health Organization* decision, which overturned *Roe v. Wade.* The achievements of the women's movement have been eroded, upended, and even obliterated. Now hardly seems the appropriate time to critique the history of Planned Parenthood or focus on the racism of progressive political history.

But Lorde's injunction shows us how the desire for social progress and a willingness to use the power of science can turn communities into populations to be measured, managed, and controlled. We are better able to recognize the violence enacted by calls for improved public health, alleviations to poverty, or the regulation of immigration when we recognize their eugenic undertones. Remembering her declaration helps us to reckon with the intermingled histories of women's liberation and white supremacy in the United States.

It is on this last point that the lens of biological determinism is most limiting. The academic critique of science that emerged from the radical science era focused on the denaturalizing of scientific knowledge. By rejecting the liberal claim to scientific objectivity, these academics highlighted the bias inherent in scientific practices and pushed against biological determinism. But when we view the history of scientific oppression through the framework of biological determinism, we tend to treat gender, race, sexuality, and class as analogous identities. We imagine social hierarchies created by elite white masculinity in which women, the poor, and minoritized racial groups are fixed in place on the lower rungs of the social ladder. We treat these categorizations as additive and therefore interchangeable. (Importantly, this was the rhetorical form used by the SSG, Gould, and the Genes and Gender Collective in their criticisms of sociobiology.) And in doing so, we make it possible to fashion critiques that only contend with one or another of these identities. But we need more than the critique of bias if we are to fully contend with the progressive nature of scientific epistemology. We must be able to see the specific and historically contingent realities created by intersecting oppressions. We need a broader vision of what a truly inclusive science and medicine look like.

Before the Civil War, biology as destiny for Black women did not mean that they were relegated to domesticity but rather that they were treated as bodies for gynecological experiments and as capital investments to be bred for a labor force.[129] Through the twentieth century, state agencies classified American Indians as disabled in order to render them wards of the state and to dispossess them of their lands and their children.[130] Today Mexicans are surveilled as always already migrants in the United States to render them outside the fold of citizenship and beyond the equal protection of law.[131] The framework of biological determinism does not provide us what we need to reckon with these oppressions or to imagine fully what liberation and flourishing look like in the face of them.

These are by no means novel insights. These were the realities that Lorde voiced in her 1979 lecture when she invoked the importance of lesbian and Third World women consciousness. They were spoken by the Combahee River Collective in 1977 when they claimed that "the major systems of oppression are interlocking."[132] They were given a vocabulary for feminist and critical race studies by Kimberlé Crenshaw's powerful analysis of the intersection of race and sex in

1989.[133] And they were animated by the political and organizing energy of the reproductive justice movement in the early 1990, and the ongoing work of groups like SisterSong and Spark Reproductive Justice Now.[134] Here I ask only how these insights reframe our approach to the history of scientific violence and shape our understanding of how a democratic science might redress those histories.

Ultimately, Lorde turns our attention to what were the most profoundly democratic approaches to science and medicine from this era, namely, the populist work of leftist health movements. These communities took the practical realities of health into their own hands by wrestling it out of the control of credentialed doctors and hospital officials. In the early 1970s, the Young Lords Party went door to door in El Barrio, in East Harlem, to test for tuberculosis, a disease eliminated among the rich and middle class but still spreading among Puerto Rican and Black communities.[135] At the same time, the Black Panther Party routinely discussed medical cases in their party publications, identifying pathologies as well as social determinants of health in their analyses.[136] And from 1969 to 1973 a network of "Janes" created by the Women's Liberation Union helped women secure safe, illegal abortions, at first connecting women to willing practitioners and then performing almost all the abortions themselves.[137] These are just a few examples of New Left political movements organizing health services, access, and medical procedures for their communities. These self-help power politics emerged out of a resounding critique of the intersecting oppressions of American imperialism, capitalism, and racism.

These leftist health movements did more than criticize the historical racism of Western science and medicine. They rejected systems of professional training and the social deference to medical and scientific expertise. Theirs was a far more radical vision than expressed by Gould in *The Mismeasure of Man*. This was not better science done through self-reflection on bias by credentialed academics. It was a complete surrender of the power of science to communities. Here is where the greatest liberation from the era of New Left criticisms of science and medicine lies.

But inherent in this promise is the loss of control. After all, populism was not, and is not, restricted to the Left. As we will see in succeeding chapters, the religious Right also organized grassroots movements to criticize the expertise of mainstream science. Creationism created alternative institutions and systems, establishing its own museums, colleges, and scientific journals. Here also a community took science into its own hands. And it did so in order to imagine a Christian American nationalism predicated on heterosexuality, patriarchy, and whiteness.

In the face of this form of right-wing populism, scientific liberalism has seemed to many of the academic Left as a necessary bulwark against ethnonationalism and demagoguery.

I opened the story of the criticisms of *Sociobiology* by calling it a flash point. But perhaps this is the wrong metaphor because it gives the impression that the political viewpoints of those involved in the controversy were fully realized in the moment of conflict. It would be better to understand the sociobiology episode as a crucible, out of which several political standpoints were honed and clarified. Gould's writing participated in the emergence of an approach to science across the academic Left that emphasized the social character of science and cultivated a suspicion of claims to scientific objectivity. Hubbard and Lowe's work helped lay the groundwork for the rise of feminist accounts of science, which put identity at the heart of historical and sociological analysis. And Wilson's defensive position in the face of the moralized shaming from the academic Left made him hesitant to publicly discuss race difference until the end of his career.[138] It also shaped his scientific liberalism, helping to create a cultural narrative about the Enlightenment that would promote a racialized depiction of science as the apex of Western culture.

Scientific liberalism and the ideas of sociobiology remained tightly intertwined from 1975 onward, cementing sociobiology's cultural identification with science. And this identification was aided by the capacity for whiteness and masculinity to claim scientific authority, regardless of the form of empirical practice or credentialing involved. In other words, sociobiology was very successful in creating a broad cultural intuition that it was *the* scientific approach to the evolution of human nature and, furthermore, that those that criticized it came from outside science. To understand the consequences of sociobiology's success, we must turn more fully to those ideas, especially ideas about the dynamics of sex difference and sexual conflict in its evolutionary vision of humanity.

CHAPTER THREE

Evolutionarily Engineered Sexism

> At its worst, the adaptationist program represents a kind of vulgar Panglossianism that relates everything we do to a material notion of utility. But the alternative view that I advocate here, with its focus on nonadaptations, frees us from the need to interpret all our basic skills as definite adaptations for an explicit purpose.
>
> —*"The Evolutionary Biology of Constraint," 1980*

The Basilica di San Marco in Venice has been famous for many things over its long history—as a center of Catholic power, a sign of civic pride, or a demonstration of the wealth of the Venetian ruling class. But in 1979 the basilica gained notoriety from a rather unexpected source when Stephen Jay Gould used its architecture as a metaphor to criticize sociobiology. Coauthored with Richard Lewontin, his article "The Spandrels of San Marco and the Panglossian Paradigm: A Critique of the Adaptationist Programme" was first published in the *Proceedings of the Royal Society of London* and quickly became one of the most discussed—and most infamous—articles in modern evolutionary thought.[1]

In "The Spandrels of San Marco," Gould argued that sociobiology was complicit in what he termed the *adaptationist program* of contemporary evolutionary biology. According to Gould, *adaptationism* was the view that natural selection was "the sole directing force" in evolution, which meant that organisms were designed by selection to have an "optimal form in an engineer's sense."[2] By contrast, Gould argued that the existing structure of organisms could constrain the free action of natural selection. He borrowed the term *spandrel* from architecture (spandrels are the roughly triangular spaces between an arch and a rectangular frame, or where two adjacent arches meet a ceiling) to describe biological features that are not an engineered solution to an environmental problem but are instead a happenstance of history. Pointing to the mosaic design on the spandrels within

St. Mark's Basilica in Venice, Gould acknowledged that they were "so elaborate, harmonious, and purposeful" that it was "tempt[ing] to view it as the starting point of any analysis" and "the cause in some sense of the surrounding architecture." But we would ultimately be wrong, Gould argued, to interpret the historical origin of spandrels as synonymous with their present use as decorative elements in the building's mosaic. Similarly, the complex and seemingly purposeful design of organismal anatomy might lead us to assume that creatures were fashioned by natural selection expressly and only for their contemporary uses. But this would also, in Gould's reasoning, "invert the proper path of analysis."[3]

To Gould it was this confidence in natural selection's engineering capacity that gave sociobiologists (and the publics they wrote for) a Panglossian understanding of the world. He took the term from the character Dr. Pangloss in the eighteenth-century French novel *Candide*, by the philosopher Voltaire. According to Pangloss, since God is good and all powerful, he must have made our world the best one out of all conceivable possibilities. All features of our world, no matter their apparent evil, are thus logically necessary.[4] By describing adaptationism as "Panglossian," Gould accused his fellow evolutionary biologists of having a similar commitment to the logical necessity of the world as it is.

A world impervious to change is a world that cannot be reformed. Even as the women's movement continued to challenge the traditional place of women in American society, sociobiologists began to argue that sex differences were optimal design solutions governed by selection. Sociobiology ushered in a new evolutionary narrative that portrayed contemporary gender roles in the United States as determined by prehistoric adaptations encoded in humanity's genes. Men, they claimed, were predisposed to the workplace; women, to domesticity. By the early 1990s another new subdiscipline, evolutionary psychology, had reinforced this narrative by extending these claims about prehistoric genetic coding to the realm of human psychology. The images that these fields built—that males are promiscuous and risktakers, while females are coy and nurturing—circulated in ongoing debates over diet and sexual health, gender disparity in STEM fields, and the root causes of sexual assault.

Beginning with "The Spandrels of San Marco," Gould launched a decades-long campaign against sociobiology's model of evolutionary causation and history. Although distributed across dozens of technical and popular texts, his campaign relied on three primary arguments. First, Gould contended that natural selection did not work on infinitely pliable starting material. In his view, existing anatomical features could direct the pathway of evolutionary change, meaning that the current makeup of an organism might be as attributable to the happenstance of history as it was to the engineering capacity of selection. Second, he argued that selection did not only work at the level of genes or even of individual organisms. Rather, natural selection acted at different taxonomic levels and on different timescales. The selection of species, for instance, could potentially minimize the importance of adaptations at the organismal level.[5] Finally, he insisted that evolutionary theory must account for truly unpredictable events in history—for catastrophes such as meteors—that dramatically affected the success of evolutionary lineages. The role of historical contingency in evolution would force us to recognize, in Gould's reckoning, that the dinosaurs went extinct because a meteor had struck the earth at the end of the Cretaceous, not because mammals were a more advanced taxonomic group.

Gould's criticisms of adaptive explanations were leveled not only at sociobiology but at the wider field of evolutionary science. In his effort to root out the reliance on selection in evolutionary thought, Gould employed sweeping rhetoric that accused his colleagues of a hopelessly limited understanding of Darwinism. And he saw the consequences of this limitation everywhere he looked. It was reflected in paleontologists' inability to see gaps in the fossil record as actual evidence about how evolution worked. It was evident in sociobiologists' insistence that natural selection only operated on genes, as well as in evolutionary biologists' refusal to consider how the preexisting shapes of organisms limited evolution. Gould claimed that a wealth of new research in paleontology, molecular genetics, and embryology rendered this version of Darwinism "effectively dead."[6] He called for a "new and general theory" that would see "the control of evolution not only by selection, but equally by constraints of history, development, and architecture."[7]

Gould's vision went beyond new research programs and disciplinary formations. He claimed that the desire to see natural selection as an agent of purpose and progress reflected "all the deep prejudices of Western thought."[8] And he attributed the scientific belief in human racial hierarchies to the pervasive metaphor of progress in evolutionary theory. When he first began writing for public audiences, at the end of the Vietnam War, in the breath between the civil rights and women's movements, Gould found this view morally objectionable. How could America have a better future if scientists claimed that the injustices of the present resulted from the inevitable logic of evolution? One solution was to set Darwinism aside as a bankrupt, sexist, and racist philosophy. But Gould did not do this. Instead, he argued that the perversions of evolutionary theory could be rectified by revisiting Darwin's original spirit. By remembering Darwin's instruction to "never say higher or lower," humans would see that evolution was not a progressive tree but an "arborescent bush" on which they were only a "tiny little twig" and thus a "fortuitous cosmic afterthought."[9] Gould contended that disrupting progressive evolutionary metaphors would force humans to admit that they were "not the loftiest products of a preordained process" and free them to embark on a new ethical project, "for if we cannot find purpose in nature, we will have to define it for ourselves."[10]

Despite the intensity of Gould's rhetoric, his criticisms of sociobiology's evolutionary models were overshadowed by the more explicitly political critiques emerging from the pens and actions of other figures in feminist and radical science groups. Nevertheless, this chapter focuses on Gould's technical criticisms of sociobiology in some detail. The sociobiological paradigm of human evolution continues to wield significant influence in American public culture. Its depiction of the qualities of masculinity and femininity continues to resonate in a wide range of public conversations, from liberal criticisms of trans identity to the devotional literature of evangelical Christianity. By rendering sex difference as the core driver of human evolution, sociobiology helped build a scientific foundation for a cultural insistence on gender as a biological binary. And this conviction of the biological fixity of maleness and femaleness continues to be leveled by both liberals and religious conservatives against queer and trans advocacy for civil rights. Understanding the technicalities of those arguments made by sociobiology—as well as Gould's contingency-based alternatives—provides us a vital lens through which to consider the insertions of biological authority into ongoing debates over gender and sexuality in American society. Showing that the world could have been different makes it possible to imagine a new and more just future. And Gould was convinced that transforming the reigning evolutionary paradigm from engineering adaptation to historical contingency would open the way for political and social equality.

Optimizing Sex Differences

One week in August 1977, the cover of *Time* magazine featured a white man and woman with their arms around each other suspended from strings like marionettes. The words "Why You Do What You Do; Sociobiology: A New Theory of Behavior" hung over the image.[11] The cover story opened with the most striking claim of the new scientific field: that "human behavior is genetically based, the results of millions of years of evolution."[12] In broad strokes, the piece conveyed to *Time*'s readers the new sociobiological evolutionary narrative. That narrative asked the American public to imagine that their genes were akin to computer programs, controlling behavior even without their consciously realizing that they were doing so. According to the article, sociobiologists described genes as having an immortal, timeless character that connected the present and the past. The opening declared that "morality and justice, far from being the triumphant product of human progress, evolved from man's animal past, and are securely rooted in the genes."[13] With this, *Time* implied to its readers that sociobiologists doubted the possibility of significant social reform in the face of the deep evolutionary logic of humans' genetic programming.

A belief in the power of DNA to explain human social traits circulated in many contexts beyond evolutionary biology in this period.[14] From the mid-twentieth century on, scientists, politicians, journalists, and reformists in many corners of American political life looked to the language of genes to explain everything from criminality, health, and intelligence to educational outcomes. But there were specific aspects of the evolutionary reasoning in sociobiological texts that gave strength to a developing cultural intuition that humans' evolutionary past was indelibly etched into their genetic code. Moreover, sociobiology was notable in its emphasis on sex difference as a core mechanism of human evolutionary development. And although sociobiology did not invent its sexual stereotypes wholesale (a fact that feminists were quick to point out), aspects of its model of sex difference were genuinely novel. By century's end, the sociobiological account had had enormous success in claiming to be *the* scientific account of gender roles.

Popular coverage of the sociobiology controversy (including the *Time* cover story) tended to characterize sociobiology as a "new" evolutionary approach to human behavior created by E. O. Wilson. In reality, Wilson's *Sociobiology* both synthesized earlier work in animal behavioral studies and catalyzed a new surge of interest in the biology of sex difference.[15] A number of volumes published soon after 1975 captured the energy generated by Wilson's book.[16] The scientists producing this work incorporated diverse methodologies and assumptions, particularly in the fields of evolutionary genetics, ecology, and behavioral anthropology. I nevertheless shorthand all of this as *sociobiology*, as the term captures the essentials of a key set of texts that influenced this scientific depiction of human sex difference.

Even before Wilson's book, a critical paper was published by the evolutionary biologist Robert Trivers. During the early 1970s, Trivers was a junior faculty member in Harvard's anthropology department working on a series of papers that would form the theoretical core of evolutionary research on sexual selection for the ensuing decades. While at Harvard, Trivers worked with Wilson, and his theories influenced the final draft of *Sociobiology*.[17] In his 1972 paper "Parental Investment and Sexual Selection," Trivers proposed that males and females were distinguished by their differential resource investment in their offspring.[18] His "parental investment" theory provided a genetic foundation for the image of sexually aggressive males and sexually coy females.

Wilson had taken up Trivers's ideas in the famous last chapter of *Sociobiology* and expanded on them in a full-length treatment of the evolution of human sociality, *On Human Nature* (1978). In *On Human Nature,* Wilson devoted chapters to the evolution of notable features of human nature, including aggression, religion, altruism, and, importantly, sex. Commenting on the last, he acknowledged that it was vital that society determine to what degree heredity and environment each contributed to the "differentiation of behavioral roles between the sexes."[19] He concluded that the genetic differences between the sexes were "modest," but nevertheless he argued that many of the essential sexual patterns, particularly in relation to roles within the family, were unlikely to be transformed by cultural efforts.[20]

Whether talking about gender roles or other social behaviors, sociobiologists tended to speak of humans and other animals as data processors. This intuition largely arose from the field's combination of population genetics with an understanding of heredity as an information system. This intuition was given a powerful metaphor when Dawkins published *The Selfish Gene* in 1976.[21] At the time, Dawkins was an ethologist and a fellow at New College, Oxford; he would later rise to international prominence as a critic of American creationism and a leader of the New Atheism movement. In *The Selfish Gene* Dawkins argued that genes were network strategists that were "trying to get more numerous in the gene pool" by "helping to program the bodies in which it finds itself to survive and to reproduce."[22] With this programming imagery, *The Selfish Gene* clearly defined the gene-centered view of evolution. Although Dawkins introduced neither a novel set of empirical tools nor statistical models, his book crystallized this enormously powerful and productive research agenda for a generation of scientists.

Sociobiologists were drawn to models of gene-centered evolution because of their ability to explain the evolution of social behaviors, which seemed puzzling at the individual level. One important example involved accounting for the evolution of altruism, that is, of behaviors that involve a cost to self to help another. Earlier evolutionary theorists had sought to explain this phenomenon through a form of natural selection that acted upon groups, the idea being that groups could somehow collude for their own greater good.[23] But new models in population genetics and game theory suggested that group-centered behaviors would always be selected against, since individuals could benefit by cheating the system.

Sociobiologists proposed that thinking of evolution as working on behalf of genes, rather than individual organisms, offered a biological explanation for seemingly self-destructive behaviors. In the model of kin selection, altruism on behalf of close relatives could be explained by the amount of genetic material they shared. If caring for a close relative harmed the body carrying the gene, the genes they shared would survive to perpetuate this sacrificial behavior.

The sociobiological view of human heredity coupled its focus on the gene as the unit of selection with new approaches to genetics that had flourished in the years following World War II. In the early twentieth century, the science of heredity had been dominated by classical Mendelian genetics. Geneticists like Thomas Hunt Morgan had conceptualized genes as chromosomal markers that revealed stories about genealogy, not packets of information that managed development or behavior. Beginning in the 1930s, genetics researchers had increasingly studied the relationship between DNA and protein expression. When molecular geneticists demonstrated that mRNA acted as a conduit between the genetic script and proteins, this cemented a new language that described genes as the "code of life."[24] Sociobiologists drew upon this informational view of heredity in their accounts of human social evolution.[25] If DNA were a code, then it could be a blueprint for behavioral patterns and social systems.

The crucial final piece in sociobiology's model of evolutionary change was its approach to natural selection. Sociobiology understood selection as an optimizing force—a notion that had been gaining traction in evolutionary fields since the mid-1960s.[26] Researchers interested in the evolution of behavioral ecology—that is, in how animals acted to maximize their genetic fitness in the context of environmental problems—used optimality models to understand how selective processes molded behavior and anatomy. Optimality models provide a way of reasoning through the evolutionary process without constructing the actual history of selective action on a population of organisms. Say, for instance, that a biologist has observed an anatomical trait and wants to know whether this trait has contributed to the fitness of organisms in a population that possess it. In theory, she could attempt to find empirical evidence that the trait had once existed at lower frequencies in the population and then demonstrate that the trait had spread through the population over the course of many generations because it conferred some sort of survival advantage to its possessors.

In practice, however, examples of specific traits like this are exceedingly rare for the simple reason that they happened in the past and potentially on a timescale exceeding a human lifetime. To address this issue, many biologists turned to optimality models. If our researcher wanted to use such a model, she would first calculate the physical parameters of a good design, that is, a design that maximized fitness from an engineering standpoint. For instance, if the biologist were trying to understand whether the mandibles of a given insect species had contributed to its fitness, she might determine how long the mandibles would have to be in order to aid in catching prey without impeding the insect's movements. With a model for optimal design in hand, the researcher would then examine the organism in question. If such a design were confirmed, its existence was understood to be an adaptation constructed by natural selection.

For some biologists, optimality models became a shortcut for evidence of the actual history of the evolutionary process. As Wilson described them in his 1978 work *Caste and Ecology in the Social Insects,* optimality models were a means of "making educated guesses as to how evolution might have proceeded." They were a kind of research heuristic, Wilson explained, that could "[suggest] avenues for further empirical research."[27] In the day-to-day practice of optimality modeling, however, this research heuristic could become a replacement for a historical, empirical account of the evolutionary history of an adaptation.

Although the field applied the optimizing model of evolutionary change to a range of human social behaviors, sociobiologists focused particularly on what they referred to as the quandary of sexual reproduction. As Dawkins put it, "What is the good of sex?" Although it appeared to be a simple question, sociobiologists contended that it was "extremely difficult" to account for the evolution of sexual reproduction through traditional interpretations of natural selection.[28] Sex appeared to be a very inefficient way for an individual to perpetuate her genes since her offspring would only contain half of her genetic material. Envisioning sexual reproduction as a battle between two partners, Dawkins and other sociobiologists argued that maleness and femaleness had evolved as two different genetic strategies for maximizing the genetic return from offspring. According to sociobiology, females employed a strategy that relied on heavily investing bodily resources into one (or very few) offspring. This made each reproductive act a high-stakes event for females, which meant that they were sexually choosy, or coy, in their interactions with males. Males, by contrast, tended to invest resources in as many offspring as possible, which was translated by sociobiologists and the media as a declaration that evolution had adapted human males to be promiscuous and to resist sexual fidelity to any particular female.[29]

These gender models pervaded professional research and circulated in the popular press, illustrated by images like the marionette couple on the cover of *Time* in 1977 Many sociobiologists, including Wilson, argued that the study of human sex difference should be regarded in the same dispassionate manner as any other field of science. But in a nation where the self-help ethos of the women's health movement demanded radical self-determination and new social possibilities for women, the view that gender was rooted in the puppeteering of evolution was unacceptable to many on the left. And in the wider context of the conservative campaigns for traditional gender roles, these liberal models seemed to parallel reactionary religious politics. In response, as we have seen, feminists criticized scientific neutrality as a mere guise for masculine social dominance. And in the next several decades, feminists both within and outside evolutionary biology would develop intuitions about gender that departed from the model of evolutionary change in sociobiology. Here, we will take time to dwell further on Gould's criticisms of adaptive reasoning and how he argued for an alternative model of evolutionary history that he claimed could also be a force in egalitarian politics.

Criticizing the Paradigm of Adaptationism

As discussed in chapter 2, feminist science collectives criticized the sexual stereotypes and lack of empirical documentation in the sociobiological literature. Gould's writing on racism in the history of science for *Natural History* and in *The Mismeasure of Man* was broadly sympathetic to the feminist approach to a critique of sociobiology. However, his main criticism of sociobiology concentrated instead on the evolutionary model at the heart of the sociobiological perspective. Sociobiologists were able to speak of humans as puppeteered by the algorithmic logic of their genes because they coupled optimality models of selection with an informational understanding of heredity. Gould's three technical criticisms of sociobiology focused particularly on its reliance on optimality modeling, that is, the view that selection acted "as an optimizing agent."[30]

It was through the concept of a scientific paradigm that Gould attempted to unite his political and scientific motivations when criticizing sociobiology. As we have seen, Gould's understanding of Kuhnian paradigms shaped his commitment to a philosophy of scientific knowledge that emphasized that science was theory-laden. Gould was so convinced of this that he asserted that it was through reassessment and reflection on the history of scientific theory that true scientific progress came. As early as 1977, in *Ontogeny and Phylogeny*, he argued that "facts never exist outside of theory, and imaginative theory may be even more essential than new information in yielding scientific 'progress.'"[31] This view—that it was not through the patient accumulation of empirical details but in the bold reassessment of scientific frameworks that advanced science—shaped both Gould's public and his technical writing.[32] While the precise reasons for

Gould's rejection of adaptive reasoning shifted over the course of his career, what is clear is Gould's desire to use theory to drive a revolution in evolutionary science.[33] He wanted to be both an ethical scientist and a figure renowned for greatness in the pantheon of evolutionary scientists. In proposing a Kuhnian paradigm shift, Gould attempted to be both.

Gould began his campaign in the fall of 1978, when he presented "The Spandrels of San Marco" at a symposium on adaptation at the Royal Society of London. As Gould later explained, Lewontin had been invited to the session "to present a contrary view" to a company of evolutionary theorists who were enthusiastic adaptationists.[34] When Lewontin was unable to attend the session, the organizers agreed to allow Gould to present a joint paper. Although Gould was new to critiques of adaptive models, this was a well-established theme in Lewontin's work. The section of the paper titled "The Adaptationist Programme" clearly drew from Lewontin's decades-long criticisms of adaptive explanations. Lewontin had written widely on this subject for both technical and popular audiences, and the paper added little to Lewontin's core technical criticisms of adaptive evolutionary reasoning.[35]

Given that "The Spandrels of San Marco" included neither a substantially new technical criticism nor a novel political critique, its impact across the evolutionary community must be attributed at least in part to its powerful rhetoric. Importantly, Gould later claimed that he had written "virtually all of the paper."[36] The two most charismatic elements of the essay—the proposition for spandrels and the contention that adaptationism was a Panglossian paradigm—came from Gould. It was in this paper that Gould argued that the adaptive stories of optimality models were not simply a misguided research heuristic, but an epistemology of social oppression. In so doing, he framed sociobiology as immoral. This accusation resonated across the evolutionary community, provoking responses for decades to come.

In addition to Lewontin's long-standing work on adaptation, the empirical arguments in the paper also drew on Gould's reinterpretation of his own early scientific work. In 1977 Gould had published his first scientific monograph, *Ontogeny and Phylogeny*, in which he argued that evolutionary biologists could discover new avenues for analyzing ecological adaptations in the vast array of potential relationships between individual development and phylogenetic change. Classic studies of species diversification, known as adaptive radiation, focused on the emergence of novel anatomical structures that allowed organisms to exploit specific ecological resources. For instance, in the case of Darwin's Galapagos mockingbirds, differences in beak shape and size allowed variants to specialize their feeding patterns, and these differences eventually produced new species. This model assumed the presence of a beneficial and heritable variation for natural selection to work upon. Gould, by contrast, suggested that variation need not come from novel genetic mutations but could instead arise from the plasticity of developmental timing. *Ontogeny and Phylogeny* inspired a generation of researchers to examine development pathways as the source for niche specialization and speciation.[37]

Although *Ontogeny and Phylogeny* contained elements of Gould's later criticisms of adaptive explanations, the volume was missing a final critical piece. The primary purpose of *Ontogeny and Phylogeny* was to argue that the relationship between development and phylogenetic change provided the raw material for ecological adaptation. The text, in other words, relied on an adaptive framework.[38] The volume incorporated the parameter of time, specifically what Gould referred to as "acceleration and retardation," into adaptive explanations, not in opposition to them.[39] Later on, Gould would attest that the book was the final time his work "stuck closely to the selectionist orthodoxy."[40] Only two short years later, "The Spandrels of San Marco" leveled a vociferous critique against adaptive explanations in contemporary evolutionary research. Moreover, Gould used the article to reinterpret his earlier work on the role of development and structural form in evolution as an alternative possibility for evolutionary explanations rather than a generative area for research on ecological adaptations.

Why did Gould's thinking about adaptation change so rapidly? Near the end of his career, Gould himself identified several explanations. He pointed to the work he had done with other paleontologists during the early 1970s and their attempts to "develop more rigorous criteria" for identifying patterns in the fossil record "that required selectionist rather than stochastic" accounts. He also cited the literature he had absorbed while preparing the section of *Ontogeny and Phylogeny* on nineteenth-century German structuralist biology. Musing on this period, Gould also acknowledged that his "growing unhappiness" with the "speculative character of many adaptationist scenarios" had increased in response to the growing literature in sociobiology that "touted conclusions that struck [him] as implausible" and also ran counter to his "political and social beliefs." At least in retrospect, however, Gould was extremely reluctant to attribute his change of heart solely to "personal distaste."[41] During the early years of the sociobiology debate, other evolutionists accused Gould and Lewontin of trumping up a technical critique of adaptive models because of their political objection to sociobiology.[42] Several decades later, the sociologist Ullrica Segerstråle made a similar accusation in her ethnographic study of the sociobiology controversy, *Defenders of the*

Truth (2000).[43] In light of these criticisms, Gould wanted to establish that he had had preexisting technical objections to adaptive reasoning that were not solely motivated by his distaste for sociobiology.

Gould's second and third criticisms of adaptationism were both arguments that history ought to matter in models of evolutionary causation. When sociobiologists incorporated optimality modeling into their work, they typically assumed that the empirical evidence of evolutionary history was immaterial to understanding evolutionary adaptations. In this they had implicitly excluded paleontology, a discipline primarily occupied with the evidence of evolutionary history, from "the high table of evolutionary theory."[44] Perhaps unsurprisingly, Gould's technical criticisms of sociobiology were also a brief on behalf of paleontology's relevance for understanding the evolutionary process. Although Gould worked in the same building as Wilson, his own view of the evolutionary process was formed by a different set of problematics than were the views of researchers interested in the evolution of behavior.

By the publication of "The Spandrels of San Marco" in 1979, Gould had been working for almost a decade with a group of other young paleontologists to revolutionize evolutionary theory through their novel approaches to fossil data. A notable aspect of this work was put together in 1972 by Thomas Schopf, David Raup, Jack Sepkoski (one of Gould's graduate students), Gould, and others in what became known as the MBL model, named for the Woods Hole Marine Biological Laboratory on Cape Code in Massachusetts. During their meeting, the group worked together in an effort to radicalize paleontology into something that would compel the attention of other evolutionary fields. Together they experimented with a statistical model of paleontological data that they hoped would allow them to capture the underlying reasons for the major taxonomic patterns in the fossil record. But after working on the model for several days with little success, Raup, frustrated, suggested that they take natural selection out of the equation. As Raup later explained, the idea was to set aside the specific causes of extinction because "there are so many different causes that ensembles of extinction may behave *as if* governed by chance alone."[45] The MBL model helped establish the careers of these scientists, but it also contributed to the formation of paleobiology as a new subfield of paleontology.[46]

Characteristically, because he was conversant in the history of his own discipline, Gould highlighted the philosophical importance of the MBL model's embrace of law-like processes in a letter to Schopf in 1973. He remarked that the paleobiologists "are among the first group of scientists to use a consciously nomothetic approach to paleontology," which Gould described as a "heretical" move in a field that was "so deeply committed to historicity."[47] What did Gould mean by contrasting the MBL model's "nomothetic" approach with the field's commitment to "historicity"? *Nomothetic* means law-like, or possessing the quality of having laws. In Gould's time and before, the physical sciences were understood to be particularly good examples of the lawfulness of science. Newton's laws, for example, were understood to explain the behavior of all particles, regardless of their place in the universe or the perspective of the observer. If paleontology could be nomothetic, then it escaped being merely descriptive of the historical results of evolutionary mechanisms that were the intellectual purview of other scientific disciplines.

The MBL model also had explicit political resonances. If history was shaped by stochastic forces as random as the motion of gas molecules, then evolutionary taxa succeeded not because of their inherent superiority but because of chance. Some in the MBL group felt that this model of evolutionary history could also reflect a truth about human history.[48] The relative success of individuals, racial groups, or genders, might also be the work of random forces. The model suggested that patterns in human history might also be the work of chance rather than the result of evolutionary competition.

The MBL model was the most radical theoretical proposition put forward by this group of young paleontologists. But it was Gould and Niles Eldredge's theory of punctuated equilibria that became the centerpiece of the paleobiological revolution.[49] Punctuated equilibria proposed a new interpretation of the evidence of the fossil record and a new picture of the tempo of evolutionary change. The traditional model of Darwinian gradualism expects the fossil record to show series of closely gradated forms that reflect the incremental work of natural selection. Instead, the fossil record overwhelmingly shows species that remain the same for long periods of time and then abruptly change. Darwin noted this problem when he originally proposed his theory of natural selection.[50] His solution was to propose that incomplete fossilization had left gaps in the record. That is, the transitional forms ought to be there, but the fossils are simply missing. In "Punctuated Equilibria: An Alternative to Phyletic Gradualism," a chapter in the 1972 book *Models in Paleobiology*, Gould and Eldredge argued that the gaps in the fossil record were not missing data but actual evidence.[51] They accordingly proposed that evolutionary change was best understood as happening in rapid bursts. A species would remain unchanged for most of its existence, but then a speciation event (such as a radically altered environment or a new geographic barrier) would cause the population to split and form new species. Punctuated equilibria shifted evolutionary explanations away from the constant work of natural selection to form adaptations, urging that evolutionists should instead focus their attention on these moments of sudden speciation.

In the decades following the initial publication of their theory, Eldredge and Gould argued that punctuated equilibria were proof that the species level was one of the levels at which selection took place. If species were real entities, then selection could act upon them and sort them for extinction.[52] And species could be selected for reasons that had little to do with their adaptive complexity. In the geologically instantaneous moment of speciation, species might be sorted on the basis of properties of the reproductive community rather than the fitness of individuals.

Punctuated equilibria posed a direct challenge to sociobiology's assumption that all trends in the history of life could be extrapolated from adaptations produced at the genetic level. One of sociobiology's central concerns was to explain the complex nature of ecological behaviors and anatomical structures. In this evolutionary framework, natural selection was the only force that generated adaptations or split lineages into new species, and it did so by acting continuously to select genes. As a field, sociobiology was much more concerned with, say, how the vertebrate eye evolved to process light than with why mammals dominated the Cenozoic era. If sociobiologists believed that optimality models were powerful shortcuts to a compelling evolutionary explanation, then paleobiologists wanted to demonstrate that the empirical realities of the fossil record disrupted the assumptions of those models.

Punctuated equilibria was an argument for the importance of history in evolutionary explanations. However, it relied on an understanding of history that was characterized not by directionality but by the scale and scope of time as a parameter in evolutionary explanations. After all, in the early days of the MBL-model group, Gould exulted that the quest for a nomothetic paleontology had led them away from the discipline's commitment to historicity. Not until the late 1980s would Gould argue that a resolute historicity was the final argument against the model of optimized selection so prevalent in sociobiology.

Gould's turn to this resolute historicity was partly influenced by new research on catastrophic extinction events in the earth's history. In 1980 a team of researchers led by the experimental physicist Luis Walter Alvarez published a study proposing an extraterrestrial cause for the extinction at the end of Cretaceous.[53] Alvarez et al. claimed that an asteroid had caused dinosaurs to go extinct. Gould read the paper with eager interest, and it influenced his view of the nature of historical change. In a 1985 paper titled "The Paradox of the First Tier: An Agenda for Paleobiology" Gould argued that mass extinctions were the "third tier" of life's history. Since they were a "recurring process now recognized as more frequent, more rapid, more intense, and more different than we had imagined" and they worked "by different rules," they might "undo whatever the lower tiers had accumulated." We might expect there to be progress in life's history because of the competition between organisms at the "first tier." However, Gould suggested that whatever progress that tier accumulated was "sufficiently reversed, undone, or overridden by processes of the higher tiers." Punctuated equilibria described the sorting of species at the second tier for "reasons unrelated to the adaptive benefits of organisms." This new, final catastrophic tier would potentially obliterate progress on the lower level.[54] By the end of his career, Gould's understanding of evolutionary history combined both timeless mechanisms and specific contingencies.

Gould's view of the nature of history was most clearly articulated in his 1987 book *Time's Arrow, Time's Cycle: Myth and Metaphor in the Discovery of Geological Time*.[55] The book was Gould's version of the history of the discovery of deep time, as told through the works of three great figures in English geology, Thomas Burnet, James Hutton, and Charles Lyell.[56] The volume staked a claim for the disciplinary importance of historical geology as well as for the validity of historical reasoning in evolutionary theory. In it, Gould argued that history comprised two opposing components. First, there was "time's arrow," in which history was "an irreversible sequence of unrepeated events. Each moment occupies its own distinct position in a temporal series, and all moments, considered in proper sequence, tell a story of linked events moving in a direction."[57] Second, there was "time's cycle," in which events had "no meaning as distinct episodes with causal impact upon a contingent history. . . . Apparent motions are parts of repeating cycles, and differences of the past will be realities of the future. Time has no direction."[58] Gould contended that in order to understand the full complexity of evolution we must recognize that both elements governed the patterns of earth's history. And he pointed to catastrophic extinction events as the most dramatic example of time's arrow.

Gould insisted that the unique particularities of time's arrow must affect our understanding of the nature of both human and evolutionary history. Further, he asserted, this sense of history had direct moral consequences. He made the connection between social ethics and time's arrow even more clearly in *Wonderful Life* (1989), a study of the historical interpretations of the fossilized fauna in the Burgess Shale, a formation in the Canadian Rockies. In the book, Gould claimed that the initial interpretation of the Burgess Shale had assumed that the fossilized forms would be simpler versions of contemporary organisms. In his view, this assumption had been driven by the persistent belief that evolution was a progressive ladder of ascending complexity. Gould triumphantly touted the later interpretations of the strange Cambrian forms, which suggested that the Burgess Shale had fossilized organisms with no present-day corollaries. In his view, once we recognized the disjunction between these early taxa and contemporary life forms, we would have to acknowledge that "life was a copiously branching bush, continually pruned by the grim reaper of extinction."[59] In other words, we would have to throw our faith in progress out the window. And if we could not understand evolution as an inevitable climb up a ladder of complexity and perfection, then we could not understand ourselves as its triumphant apex. In the book's best-known metaphor, Gould declared that if the tape of life could be replayed from the beginning, it was unlikely that anything "like human intelligence would grace the replay."[60]

When Gould published *Wonderful Life*, he sent copies to various scientific and political dignitaries. One of the recipients of *Wonderful Life* was former President Jimmy Carter. Carter wrote back to Gould after reading the book in the winter of 1989. He said that he had found it to be thoroughly enjoyable, "perhaps your best so far." But Carter questioned Gould's claim that "if the tape of life was replayed in countless different ways it is unlikely that cognitive creatures would have been created or evolved." If the odds were so low, Carter suggested, might it not be the case that "a creator" "has done some of the orchestrating"?[61] Carter's reaction to the book was a common one. Immediately after its publication and for ensuing decades, the book was interpreted as taking part in debates over chance and the divine origin of life.[62] In his response to Carter, Gould affirmed that the question whether "human evolution [can] become so improbable by the argument of contingency that its very occurrence demands a principle of design" was "absolutely the heart of the issue."[63] But Gould maintained that he had not meant for his argument about replaying life's tape to suggest that the improbability of humanity required the hand of a creator; rather that it should destroy our faith in evolutionary progress. By emphasizing the inherent importance of history in evolutionary explanations, Gould wanted to overturn the emphasis on adaptive explanations in sociobiology and the wider fields of evolutionary science. He believed that the sociobiological view of evolutionary history was simply a reconstruction of the early modern Great Chain of Being. He was convinced that understanding selection as a force of optimization had led to visualizing evolution as a ladder of progress, which he argued had inevitably led to racism.

During the period of these arguments Gould went from a relatively unknown junior faculty member at Harvard to an internationally published and renowned public scientific celebrity. His increased fame prompted resentment from some colleagues who believed that Gould's media platform had expanded the audience for his distinctive view of Darwinian evolution.[64] But even with this popular influence, Gould remained focused on transforming his scientific colleagues' understanding of evolutionary theory. He insisted that the current evolutionary paradigm—which he characterized as both morally suspect and an impediment to the advancement of scientific knowledge—could be forced to change through the collective self-reflection of the evolutionary community. And Gould was convinced he could push his fellow evolutionary biologists toward this change through a thunder of rhetoric, castigations, and provocations.

Adaptive Sexuality

In the two decades following the publication of "The Spandrels of San Marco," Gould promoted an evolutionary narrative that emphasized contingency and a lack of predictable progress. In 1996 Gould returned to the subject in *Full House: The Spread of Excellence from Plato to Darwin*, which he intended as "something" of a "companion" to the earlier *Wonderful Life*.[65] "Together," he claimed, the books presented an "unconventional view of life's history and meaning—one that forces us to reconceptualize our notion of human status within this history." Gould denied that "progress defines the history of life or even exists as a general trend at all."[66] In these and other popular writings, Gould consistently blamed the concept of evolutionary progress for providing scientific justification for racism and other social inequalities. But despite Gould's nearly unparalleled public profile, his views on the social implications of evolution, particularly for sex, were not the ones that found their way into American culture. While Gould wrote prolifically about race, he spent relatively little of his time spelling out the implications of his view of evolution for understanding the biology of human sex difference. Even if Gould's science provided some hope that sex differences were not engineered by evolution, this hope was buried in technical journal articles rather than expressed in his more effusive popular prose.

Sociobiologists, by contrast, found ready audiences for their accounts of sexually aggressive males and coy females. Cultural commentators and politicians deployed biological models of gender and sexuality in debates over sexual violence, gender equity in the workplace, and even the roles men and women played in Protestant church leadership. By the end of the century, the thread of sociobiological sexual logic would find its way into discourses about dating, rape culture, diets, exercise, education, and medicine. The most direct extension of the field came with the emergence of evolutionary psychology, which shifted the sociobiological emphasis on behavior to a model of evolutionarily adapted psychologies. Much of the literature in the field focused exclusively on the adaptive nature of sexed psychologies and of human sexuality in general.

One of the earliest sociobiological texts to concentrate specifically on human sexuality was Donald Symons's 1979 book *The Evolution of Human Sexuality*. Trained as an anthropologist, Symons built an account of human sexuality founded on Trivers's parental investment theory. In Symons's view, because eggs were a limiting resource for the male sexual appetite, men would seek to have sex with as many women as possible and treat them in a proprietary way. As he summarized his view of the sexual differences between men and women, men "incline to polygyny," "almost universally experience sexual jealousy of their mates," are "much more likely to be sexually aroused by the sight of a woman than women are by the sight of men," and "are predisposed to desire a variety of sexual partners." Symons's depiction of the masculine strong sexual appetite stood in contrast to his portrayal of women, who he argued were unlikely to be sexually jealous and who were more malleable in terms of their preferences for being "polygynous, monogamous, or polyandrous."[67]

Symons's text hinted at a development that would occur as the field grew, which was an emphasis on psychology rather than on behavior or appearance. This shift to psychology allowed evolutionary psychologists to develop adaptive explanations of sex difference, even if human anatomy or behavior offered little immediate evidence of these differences. Symons argued that "psyche—more than structure or behavior—is adapted to atypical events" and thus it could be the case that "Homo sapiens, a species with only moderate sex differences in structure, exhibits profound sex differences in psyche."[68] That is, although humans did not exhibit the clear sexual dimorphism of other vertebrates, Symons could nevertheless argue that sex differences were profound in human evolution by drawing inferences from psychology.

This turn to psychology was significant for reasons that would become clearer in evolutionary psychology texts in the following decades. For instance, Jerome H. Barkow, John Tooby, and Leda Cosmides, the editors of the 1992 volume *The Adapted Mind: Evolutionary Psychology and the Generation of Culture* wrote that the "evolved structure of the human mind is adapted to the way of life of Pleistocene hunter-gatherers, and not necessarily to our modern circumstances." The central premise of the book, which was widely seen as a foundational text for the new field, was that a universal human nature "exists primarily at the level of evolved psychological mechanisms" by "natural selection over evolutionary time."[69] The editors argued that cultural variability was simply a facade for the underlying reality of an evolved universal human nature. If true, the claim directly undercut the arguments marshaled by the Genes and Gender Collective and other feminists that cultural variability was proof against sociobiological accounts of universal sex-specific behavior in humans. Sociobiology had argued for universal patterns in gendered behavior, claims that could be disproven by demonstrating real differences between human cultures. But by arguing that humanity had been evolutionarily engineered at the level of psychology, evolutionary psychologists could dismiss cultural differences in behavior as superficial rather than as evidence against universal sex-specific traits. Moreover, this argument also allowed evolutionary psychologists to claim that contemporary society was but a shallow veneer over prehistoric human sexual psyches. In this mode, feminism could be depicted as a recent (and foolish) attempt to reform the deeply embedded realities of sexed human nature.

The field's depiction of the push and pull between men and women increased its relevance to a variety of cultural issues, including sexual violence. For instance, in their 2000 book *A Natural History of Rape: Biological Bases of Sexual Coercion*, the biologist Randy Thornhill and the anthropologist Craig T. Palmer argued that rape was a male reproductive adaptation. Thornhill and Palmer were writing after two decades of efforts by the women's health movement to build domestic violence shelters and rape crisis centers, as well as to establish laws that would recognize (and criminalize) sexual assault within marital relationships.[70] Rejecting the feminist portrayal of sexual assault as the result of patriarchal socialization, Thornhill and Palmer contended that it was the evolutionary compulsion to pass along their genes, not the desire for social power, that caused men to rape women.[71] Although Thornhill and Palmer defended their account against feminist critics, arguing that feminists had fallen prey to the naturalistic fallacy, the book generated heated controversy. According to Thornhill and Palmer, masculine sexual aggression was the natural conclusion of the male evolutionary strategy.[72]

Evolutionary psychologists depicted male homosexuality as a form of emasculation since it lacked the defining masculine desire for many female partners. They placed male homosexuality within the feminized space of domesticity. As Wilson had suggested in *Sociobiology* (and would later expand on in *On Human Nature*), "The homosexual members of primitive societies may have functioned as helpers, either while hunting in company with other men or in more domestic occupations at the dwelling sites. . . . Genes favoring homosexuality could then be sustained at a high equilibrium level by kin selection alone."[73] Since evolutionary psychology required that maleness act to maximize its genetic legacy through promiscuous copulation with females, homosexuality could only be understood as a kind of evolutionary "helper" to heterosexuality. Other texts reinforced the image of homosexuality as an aide to heterosexuality by arguing that homosexuality resulted from the evolutionary usefulness of "non-reproductive sex" to "build and maintain a network of mutually beneficial relationships."[74] The conclusion of this logic was that gay men were more in touch with feminine qualities—emotions, social niceties, and relationship dynamics.

Evolutionary psychologists coupled this image of homosexuality as emasculation with a prurient fascination with the possibilities for male sexuality if left unchecked by female sexual coyness. This fixation would continue for several decades. As Matt Ridley, a British libertarian politician and science writer, would claim in his 1993 book *The Red Queen: Sex and the Evolution of Human Nature,* male homosexuality was characterized by a greater degree of promiscuity than heterosexuality, and this was because "male homosexuals are acting out male tendencies or instincts unfettered by those of women."[75] In this and other evolutionary psychology texts, male homosexuality was treated as an unhealthy sexuality run amok without the benefit of feminine restrictions. In these texts, maleness was defined by hypersexuality that was only checked by the opposing evolutionary aims of female sexual restrictions.

Evolutionary psychologists defined masculinity by hypersexuality, even to the point of violence. Although these authors wrestled with the social consequences of male sexual aggression, they were simultaneously loath to disregard its potential power. One important reason for this was the link that evolutionary psychologists made between male sexuality and the capacity to excel in the STEM fields. Masculine hypersexuality emerged as the keystone for the crowning achievement of human evolution—science itself.

Steven Pinker, one of most prominent and commercially successful authors in the field, clearly linked masculine sexuality and science in his 2002 book *The Blank Slate: The Modern Denial of Human Nature.* Pinker argued that the relative lack of women in science and engineering fields could be the result of psychological differences between the sexes, not social forces that discouraged women from those careers.[76] According to Pinker, evolution drove "average differences in *preferences,*" including that girls were "more interested in people, 'social values,' humanitarian and altruistic goals," while boys were "more interested in things, 'theoretical values,' and abstract intellectual inquiry." This, he wrote, was the reason why "young women chose a broad range of courses in the humanities, arts, and sciences, whereas the boys were geeks who stuck to math and science." Pinker further concluded that "men are greater risk takers, and this is reflected in their career paths even when qualifications are held constant."[77] Men were naturally more fascinated by "ohms, carburetors, or quarters."[78] Women, on the other hand, were more concerned with "intimate social relationships" and "feel more empathy toward their friends."[79] In other words, because female psychology was cooperative and relational, women were less likely to choose math and science suggesting that the aptitude for these subjects was a feature of masculine competitive aggression and antisocial tendencies.

Science, by implication, needed men's sexual aggression and willingness to take risks. Moreover, science as depicted by evolutionary psychology was the province of the West, described as a contrast to the cultures of hunter-gatherer societies past and present. Unlike nineteenth-century texts, which defined Western white identity by physical features, modern evolutionary psychology rejected biological definitions of race. The writers' invocations of Westernness, however, were developed through contrast with cultural others—both within and outside the United States—who were distinctly racialized. For example, Pinker's analysis of "African-American inner-city neighborhoods" described them "as among the more conspicuously violent environments in Western democracies" because of "their entrenched culture of honor."[80] It was only through the adoption of "peaceable middle-class values," Pinker suggested, that African American families could become fully part of Western democracies.[81] And these democracies, which "almost never wage war on one another," themselves rested on a foundation of Enlightenment science.[82] White Western culture enabled the rational reflection and scientific perspective needed to rise above inherent male aggression. Ultimately, the Western self was not a biological given but a cultural accomplishment. As Dawkins stated, "Let us understand what our own selfish genes are up to, because we may then at least have the chance to upset their designs, something that no other species has ever aspired to,"[83] and evolutionary psychologists argued that it was through the maturity of scientific thought that men found their place in society.

The result of these intertwined narratives—the naturalness of male hypersexuality alongside its perceived role in formation of Western science—shaped the ongoing debates between feminist accounts of science and evolutionary psychologists. Not only did evolutionary psychologists assert that feminist accounts of evolution were misguided but they claimed that feminist criticism itself was a danger to the progress of science. In this way, a specific evolutionary psychological model of human sexuality was hitched to a more general liberal defense of the authority of Western science. And evolutionary psychology became a cultural phenomenon in the first decade of the twenty-first century, exceeding even sociobiology's explanatory powers in the 1970s. It drove the commercial success of countless books on sex and relationships, diet, and parenting.[84] It was fueled by financial investments in evolutionary research at Ivy League institutions.[85] By the early years of the new millennium, evolutionary psychology was *the* way to advocate for and do science.

The extensive reach of these models of human sexuality is perhaps best illustrated by the influence of sociobiology and evolutionary psychology in an unexpected corner of American society. During this period, the biologized model of sexually coy women and sexually aggressive males was taken up by conservative Protestant men in their efforts to defeat feminism in American culture, as well as within the evangelical community. Much of the evangelical response to the ongoing revolution in sexual values and gender roles was contained in the social campaign of Phyllis Schlafy against the Equal Rights Amendment and in the efforts of James Dobson, who founded Focus on the Family in 1977 in order to promote fundamentalist perspectives on marriage, sexual identity, and gender roles.[86] But as evangelical feminists also began to demand greater equality between men and women within the church itself, some male pastors turned to biology as an intellectual resource in their battle with feminism.

In texts such as the 1974 book *All We're Meant to Be: A Biblical Approach to Women's Liberation*, by Letha Scanzoni and Nancy Hardesty, evangelical feminists argued that the Bible supported greater gender equity in Christian homes and churches.[87] In response, evangelical men argued that feminist reform was in fact the offspring of a fallen culture. In a move that echoed a long tradition of Protestant natural theology, these pastors argued that biological science confirmed that God had designed men and women to carry out specific social roles. Men were meant for leadership, while women were designed to submit to the authority of men at home and in the church.

One of the key leaders of this conservative evangelical movement, which came to be known as complementarianism, was the American Reformed Baptist pastor John Piper, who helped found the Council on Biblical Manhood and Womanhood (CBMW) in 1987. The CBMW promoted complementarianism within evangelical churches and developed manifestoes outlining the council's stances on a range of gender and sexuality issues. Its 1987 manifesto, the Danvers Statement, argued that "distinctions in masculine and feminine roles are ordained by God as part of the created order" and that "some governing and teaching roles within the church are restricted to men."[88] In 2017 a new manifesto, the Nashville Statement, affirmed the council's conviction "that it is sinful to approve of homosexual immorality or transgenderism."[89] To support these claims about Christian sexual ethics, the CBMW drew on not only biblical exegesis and Protestant doctrine but also sociobiological models specifically in support of its claim that God had designed complementarianism and heterosexuality as part of the natural world. For instance, the CBMW's edited volume *Recovering Biblical Manhood and Womanhood* (1991) argued that differences that required masculine leadership and feminine submission were "not only real but likely have their roots in our unique biology as males and females."[90] To substantiate this point, the authors drew on evidence from sociobiological texts, including Martin Daly and Margo Wilson's 1978 volume *Sex, Evolution, and Behavior.*[91] Throughout this period, evangelical dating manuals and guides to gendered discipleship drew on biological models of maleness and femaleness to support arguments for gender complementarianism.[92] In the eyes of many conservative evangelicals, biology was a welcome ally in the culture war against feminism.

Sociobiologists and evolutionary psychologists did not intend their science to be taken up by conservative evangelical Protestants. In fact, they were more likely to view these communities antagonistically, especially once the creation science dominated national headlines in the early 1980s. Nevertheless, evangelicals and evolutionary psychologists found common ground in their shared intuition that biology offered a defense against the social reform efforts of second- and third-wave feminism. Despite their severe disagreements over Darwinism, there was a certain synergy between the logics of the two worldviews. Sociobiologists were confident in the engineering capacity of natural selection; evangelicals put their faith in the design of God. Both perspectives presented the world as sensible, purposeful, and unlikely to be changed by the paltry efforts of human culture. Many individuals in both communities portrayed feminists' attempts to remake gender roles as foolhardy at best and dangerous at worst.

The Political Promise of Historical Contingency

In the summer of 1939 the civil rights activist, sociologist, and historian W. E. B. Du Bois published an essay in the *American Scholar* titled "The Negro Scientist."[93] In it Du Bois relayed to his readers an exchange he had had with a "great American scientist" over the fact that "few Negroes had made their mark in science." Although Du Bois had called "his attention to the fact" that prejudice had made it difficult for Black Americans to pursue science and the scientist had acknowledged it, Du Bois theorized that "along with most Americans his private belief was that exact and intensive habit of mind, the rigorous mathematical logic demanded of those who would be scientists is not natural to the Negro race." What Du Bois did next in his essay was striking: he went through an exercise of "setting down the main facts" of twelve Black scientists in order to determine whether the "matter of color prejudice" or "difficulties of temperament and personality" explained why they had not fulfilled their early promise.[94] Through this exercise, Du Bois asked his readers to consider iterative examples of a historical counterfactual: would these Black scientists have achieved prominence if not for the racism they faced in early-twentieth-century America?

Du Bois's essay highlights the importance of counterfactuals in writing history: deploying them can help us to reflect on the explanations for (or even the causes of) our present-day circumstances.[95] In this case, Du Bois wished to dismiss one possible explanation for the relative lack of prominent Black scientists—their inherent or biological insufficiency—in order to promote another one—the crushing racism of their circumstances. Considering these types of counterfactuals helps us to better understand the political promise of historical contingency. What Du Bois calls our attention to is the importance of being able to imagine that the past might have been otherwise (the counterfactual) in order to imagine that our present might also have been otherwise. And it is this unpredictability of the past that is crucial for our ability to imagine alternative possible futures.[96] When Du Bois imagined that the historical experiences of those twelve Black scientists might have been different, he was also making the argument for a potential future of the nation, one in which the nation recognized the "ability in the Negro race—a great deal of unusual and extraordinary ability, undiscovered, unused and unappreciated."[97]

This relationship between a contingent past and our ability to imagine or even to create possible futures was critical to Gould's political vision in *Wonderful Life*. He acknowledged that for some, recognizing that humanity itself was a "wildly improbable evolutionary event within the realm of contingency" might be "depressing." But this was something Gould "regarded . . . as exhilarating, and a source of both freedom and consequent moral responsibility."[98] This was an argument that the very character of the history of life on earth forced us to recognize that there was no determined path before us. And if our present was not the result of the inevitable unfolding of a series of optimized steps by the engineering prowess of natural selection, then we would have to take responsibility for it, and for the future as well. With historical contingency Gould rejected both the natural theology of Voltaire's day—the theodicy that asked people in the eighteenth century to regard plagues and earthquakes as ultimately part of God's perfect design—and the adaptive apologetics of sociobiology—which asked people in the twentieth century clamoring for reform to regard social structures as part of the engineering design of natural selection.

Gould's contention that the evolution of humans was highly improbable was captured most famously by his "replaying the tape of life" metaphor, which generated enormous philosophical, historical, and biological debate from 1989 on.[99] Subsequent scholars have discussed what kind of tape Gould meant (perhaps a computer tape rather than a cassette),[100] whether his definition of contingency throughout the book was philosophically consistent (almost certainly not),[101] and whether his understanding of the Burgess Shale or of mass extinctions in the history of life was empirically sound (also almost certainly not).[102] What is clear is that Gould's "replaying the tape of life" joins Dawkins's "selfish gene" in the world of promising and contentious scientific metaphors.

Less discussed have been the implications of Gould's arguments in *Wonderful Life* for debates over racism and the history of science. This is perhaps unsurprising given that unlike *The Mismeasure of Man, Wonderful Life* was not marketed as a book about race, race science, or racism.[103] But in teasing out the political importance of historical contingency it becomes possible to regard *Wonderful Life* and *The Mismeasure of Man* as distinct but related answers to the question what should be done about racism in the history of science. As we have seen, *The Mismeasure of Man* focused on the impact of social bias on scientific practice. And it generally regarded the wider social context as the source of such bias; in the case of Morton, for instance, his racial animus was borne out of his desire to promote the American slave state. But *Wonderful Life* focuses on something different. It regards the conceptual architecture within science—in this case, a model of evolutionary causation and history—as potentially productive of racism. In this sense, it was not so much the direct discussion of racial variation or race difference in sociobiology that created racism. Rather, Gould blamed sociobiology's confidence in the engineering prowess of natural selection, an optimizing capacity that would create an image of "a conventional racist ladder of human groups up to a Caucasian paragon."[104]

It has been in reading these books alongside one another that I have found the most inspiration from Gould, because they demonstrate the power of history. In both books, the practice of history is more than a faithful accounting of past events. In *The Mismeasure of Man,* historical reflection allows us to see our past sins so that we might create less harmful—and truer—science. But in *Wonderful Life,* writing history becomes an exercise in contingent explanations of the present that may inspire us to imagine what futures might be possible. It is a call to write history to find both freedom and moral responsibility. It is an injunction to not only account for the racism of our past but also acknowledge that it is within our power to create a liberated future.

My reading of *Wonderful Life* is an unusual one. Certainly, most of the audience for Gould's book regarded it as Carter did, that is, as asking whether the evolutionary improbability of humanity meant that God was required to explain our existence. Moreover, the complex terrain of American intellectual culture during the 1980s and 1990s, a space in which leftist, liberal, and conservative religious approaches to scientific knowledge competed for space and air, often obscured the differences between leftist and liberal approaches to science. Thus far, returning to sociobiology and the political context that birthed *The Mismeasure of Man* has allowed us to isolate this debate on the very eve of the creation science controversy. But even as leftist accounts of science grew bolder after the early 1980s, they were overshadowed by another set of challenges to the authority of science that came not from activism at American universities but rather from the institutions of Protestant fundamentalism.

CHAPTER FOUR

Scientists Confront Creationism

> They are a motley collection to be sure, but their core of practical support lies with the evangelical right, and creationism is a mere stalking horse or subsidiary issue in a political program that would ban abortion, erase the political and social gains of women by reducing the vital concept to an outmoded paternalism, and reinstitute all the jingoism and distrust of learning that prepares a nation for demagoguery.
>
> —*"A Visit to Dayton," 1981*

In the winter of 1981 Gould took a trip to the small town of Dayton, Tennessee. His first wife, Deborah, and his two sons, Jesse and Ethan, accompanied him on what became the evolutionist's equivalent of a spiritual pilgrimage. Gould reflected upon the excursion in an essay titled "A Visit to Dayton" for *Natural History*. He chronicled his visits to a number of quaint local attractions, including the "well kept" older houses and the "soda-water emporium of the estimable Robinson's."[1] Along the way, he commented on some more contemporary touches, including condoms he purchased for a quarter at the vending machine of a local service station. But Gould did not journey to the South only to reflect on modernity and regional differences. He went to Dayton to retrace the events of the almost sixty-year-old Scopes (Monkey) trial. *The State of Tennessee v. John Thomas Scopes* had been a political tactic on the part of the American Civil Liberties Union (ACLU) in the summer of 1925 to strike down laws against the teaching of evolution in American biology classrooms. Gould relived the Scopes trial as preparation for a battle with the American creationist movement in the US federal district court case *McLean v. Arkansas*.

The *McLean* case arose when a group of clergy members and educators and the ACLU brought a suit against the state of Arkansas for its passage of Act 590, the

Balanced Treatment for Creation-Science and Evolution-Science Act.[2] Act 590 mandated equal time for creation science as for evolutionary accounts of what the act termed "origins" in public science classrooms. The plaintiffs in the *McLean* case argued that the motivation for teaching creation science was in fact primarily to encourage religious instruction and that therefore the bill violated the establishment clause of the First Amendment.[3] The presiding federal district judge, William Overton, eventually ruled in favor of the plaintiffs. His decision would influence the 1987 US Supreme Court case *Edwards v. Aguillard,* which ruled the teaching of creation science in public schools unconstitutional for the entire country.

In Gould's public writing, he argued that the form of creation science on trial in 1981 reanimated the same conservative politics that had driven the antievolution movement decades earlier. He insisted that "the creationists have presented not a single new fact or argument" and that the "rise of creationism is politics pure and simple."[4] Nor was Gould alone in returning to the Scopes trial as a way of making sense of the creation science movement. In the decades after *McLean,* both historians and journalists would contextualize the trial within a larger narrative that placed the court at the heart of adjudications between science and society in America across the twentieth century.[5] This framing helped solidify a broadly held intuition that the controversy over creationism and evolution was a perennial feature of American public life, peppered with a series of court cases representing legal strategies and counterstrategies between two opposing parties.

But interpreting *McLean* only through the legacy of *Scopes* obscured the actual political importance of the trial. The creation science on trial in Arkansas was integral, institutionally and intellectually, to the new Christian Right. It was not simply an assertion of Christian tradition or a mere outgrowth of fundamentalist movements. Rather, attempting to introduce creation science into American classrooms was a deliberate and essential strategy in a political movement that was gaining force in American public life. This movement wedded together evangelical theology, patriarchal and heteronormative gender values, and, more prominently in later decades, white ethnonationalism. The 1980s were far from the first moment that conservative Protestants believed the American political process had eternal consequences. Evangelicals worked throughout the twentieth century to influence state politics and education policies and to secure the place of biblical Christianity in American culture.[6] Many of these efforts explicitly challenged scientific expertise in the attempt to control public education and sexual behavior.[7] What was different in 1980 was the partisanship of these political efforts. When evangelicals threw their support behind Ronald Reagan and the Republican Party in the 1980 election, they were helping create a morally infused national political agenda that continues to influence American politics and culture in the twenty-first century.[8] Although both creation science and the larger Christian Right movement had much earlier beginnings, they coalesced into a powerful political movement in the early 1980s. In this moment evangelical religion, conservative politics, and a distrust of the scientific establishment were cemented as a cohesive package.

Despite Gould's efforts to frame *McLean* as merely a rehashing of the issues in *Scopes*, he did recognize that creation science was tied in important ways to the Christian Right. In several pieces he penned leading up to and following the McLean trial, he argued that creationism was just "one issue" amid the Christian Right's commitment to "defeating the Equal Rights amendment and a total ban on abortion."[9] In his writing for newspapers, his own column, and leftist journals, he consistently denounced creation science in the context of right-wing politics.[10] But despite Gould's assurances that creationism was simply the newest guise for conservative religion, other professional evolutionists voiced feeling a sudden onslaught in response to the newly prominent place of creationism on the national stage.

To many liberal and leftist intellectuals, it seemed that religious fanatics had built their political power, elected Ronald Reagan to the presidency, and ushered in a war against science, secularism, and liberal values.[11] The tone of surprised indignation on the part of evolutionists was mirrored throughout the political Left. The confrontation with creationism recalibrated the terms of earlier debates between liberals and leftists over the relationship between scientific authority and American democracy. During the sociobiology debate, Gould and Wilson had disagreed over the proper relationship between scientists and public dialogue, Gould contending that it was necessary to be transparent about the failures of science in order to guard against scientific racism and sexism and Wilson wanting to protect the integrity of science from the ignorance of lay interference. When creationists questioned the empirical validity of evolution, they introduced fundamentalist religion as an alternative source of authority into these public conversations. Placing creationism at the heart of the rise of the Christian Right does more than flesh out our understanding of one specific political movement. It allows us to fully locate the place of scientific expertise in one of the most dramatic political realignments in America in the twentieth century.

Creationism and the Christian Right

During the late nineteenth and early twentieth centuries, Anglo-American Protestants in the Reformed and Wesleyan traditions, who emphasized the centrality of sacred scripture, individual piety, and spreading the gospel, began to describe themselves as *evangelicals*.[12] By the early decades of the twentieth century, a different term, *fundamentalism*, was used to describe a distinctive movement within the broader evangelical tradition. Partially in reaction to modernist theological movements, fundamentalists emphasized a plain-sense adherence to the Bible, mission, and individualistic piety. Early-twentieth-century fundamentalism did not, however, attempt to discredit natural science on the national stage, set up alternative scientific institutions, or offer alternative scientific theories, not even during the Scopes trial. When fundamentalists advocated for laws that opposed the teaching of evolution in schools, they positioned their campaigns as bulwarks of traditional values against the civic aims of the urban, progressive movement.[13] Officials in Tennessee defended the law as an appropriate exercise of the right of states to establish school curricula, but they did not propose or attempt to teach alternatives to evolution, nor did they organize political action against professional scientific institutions.

Early historical work on fundamentalism argued that fundamentalists retreated from the national mainstream after their defeat at the Scopes trial. But since the 1980s, American historians have focused on the greater continuities between the fundamentalists of the early part of the century and later evangelicals. Despite a persistent caricature of fundamentalism and antievolutionism as a rural phenomenon, historians have stressed that the leaders of these movements often came from northern urban areas.[14] More recent historical work has continued to nuance the regional movement of fundamentalism to more carefully account for evangelicalism's rise in California and the Sunbelt suburbs.[15] The anthropologist T. M. Luhrmann points to the influence of hippie culture on fringe elements of Christianity during this period as one origin for an evangelical spirituality that earlier generations of Americans would have "regarded as vulgar, emotional, or even psychotic."[16] Another crucial source for this energetic spirituality was the rise of Pentecostalism during the twentieth century. Pentecostalism was a renewal movement within evangelicalism that stressed personal experience of God through the baptism of the Holy Spirit.[17] By the 1970s these influences had helped to transform conservative Christianity into something more than a theological perspective. Being an evangelical was an all-encompassing spiritual identity, one that called its followers to increase their visibility on the national scene.

But although historians have documented a persistent desire by evangelicals for national political influence, it was not until the 1980 presidential election that evangelicalism emerged as a dominant force in partisan politics.[18] Leading up to the 1980 election, a number of key Christian leaders succeeded in rallying evangelicals to the Republican Party over a number of significant issues, including abortion, gay rights, and the tax-exempt status of Christian institutions. Evangelical leaders, including Jerry Falwell, believed that the Republican Party under Ronald Reagan's leadership was the most promising place for evangelical influence and encouraged their followers to align themselves accordingly. This was a somewhat surprising decision insofar as President Jimmy Carter was an avowed evangelical and Ronald Reagan was not. However, sexual morality was such a crucially defining aspect of political evangelicalism that evangelical leaders abandoned Carter when he supported aspects of the gay rights movement and refused to back evangelical pro-life campaigns. Reagan did not fulfill many of the hopes of the evangelical Right movement, but evangelical leaders continued to support him and the Republican Party through the rest of the 1980s. By the end of the twentieth century, evangelicalism was inextricably associated with the Republican Party and its platform.

One of the most significant voices in the transforming world of twentieth-century evangelical politics was Francis Schaeffer. Schaeffer had begun his career as a Presbyterian minister, and in 1948 he moved with his wife, Edith, to Switzerland, where he establishing a spiritual community called L'Abri (The Shelter).[19] Intended as both a spiritual and philosophical center, L'Abri attracted thousands of young evangelicals, who came to debate existential philosophy with Francis and Edith. For many of the thousands of young people who flocked to L'Abri over the decades, Schaeffer and the community offered an intellectually energizing approach to evangelical identity. By the late 1970s Schaeffer had published a series of writings arguing that the legalization of abortion in the United States was a sign that Darwinian materialism had caused the West to abandon its Christian foundations.

In the summer of 1979, Francis and his son, Frank Schaeffer, embarked on a film tour that helped transform the evangelical approach to abortion. The two showed their film series, titled *Whatever Happened to the Human Race?*, to crowds in churches, in town halls, and in gymnasiums of fundamentalist colleges across the United States. The opening scene of the first film in the series, *Abortion of the Human Race*, depicts two surgical teams. One operates on a newborn who is struggling to survive. The other arranges ghastly-looking obstetric instruments on an operating table, preparing for an as-yet-unknown procedure. As the first team works, it becomes clear that they are saving the baby's life; soft music soon starts to play as the baby is transferred to a recovery room. Meanwhile, the other team prepares for a different outcome. Although an abortion is not explicitly depicted, the implication that the pregnancy is to be terminated is clear. A fully rounded pregnant abdomen appears in the frame; prepped for surgery, it is momentarily draped with a sign bearing the words "remove and discard."[20] The juxtaposition between the premature baby and the aborted fetus in the film would be echoed in arguments Schaeffer made in his 1982 book, coauthored with C. Everett Koop, the US surgeon general under Ronald Reagan. In the book, also titled *Whatever Happened to the Human Race?*, Schaeffer and Koop argued that modern society had become "schizophrenic," resuscitating infants at the same stage of development as it was legal to terminate a pregnancy.[21] The film scandalized and shocked many of its audiences; in fact many churches refused to show the film at all. But it also galvanized a generation of evangelicals to take up the antiabortion cause and helped cement abortion as one of the key political issues in the newly forged alliance between the Republican Party and the Christian Right.[22]

Although these kinds of arguments do not seem new in the early twenty-first century, at the time they represented a novel approach to the ethics of abortion. The cornerstone of Schaeffer's new theology was the argument that the practice of abortion was a sign of the decline of Christian morality and the rise of secular humanism in Western civilization. In *Whatever Happened to the Human Race?* he asked, "Why has our society changed? The answer is clear. The consensus of our society no longer rests upon a Judeo-Christian basis but a humanistic one."[23] Moreover, Schaeffer specifically blamed new evolutionary theories for the advancement of secularism. He derisively described the logic of the new sociobiological account of evolution to his readers thusly: "Regardless of what you think your reasons are for unselfishness, say the sociobiologists, in reality you are only doing what your genes know is best to keep your gene configuration alive and flourishing into the future. This happens because evolution has produced organisms that automatically follow a mathematical logic."[24] And it was this materialist conviction, the view that humans were nothing more than their genes, that Schaeffer blamed for the unfeeling and cruel nature of humanism.

Schaeffer pointed specifically to Wilson's biologized ethical framework in *Sociobiology* as an example of this depraved moral perspective. In his own writing, Schaeffer relayed Wilson's call for morality to be "biologized" and removed from "the hands of the philosophers."[25] Schaeffer was appalled by the view that "ethics and behavior" were now to be "put into the realm of the purely mechanical, where ethics reflect only genes fighting for survival." He blamed evolutionary theory as the foundation of this humanism, arguing that "if man is not made in the image of God, nothing then stands in the way of inhumanity. . . . Human life is cheapened."[26] This cheapening, Schaeffer believed, had been responsible for the Third

Reich's totalitarian disregard for life in the Final Solution. And he believed that this same moral philosophy was governing America's legalization of abortion.

Although evangelical Protestants had launched piecemeal objections to abortion policy before this period, for most of the twentieth century the antiabortion movement had been a largely Catholic cause. Protestants were not consistently opposed to eugenics, nor were they leaders in antiabortion activism.[27] When evangelical Protestants began to be involved consistently in organized antiabortion politics in the late 1970s, their presence changed the group's political identity as well as the movement itself. Schaeffer's theological arguments were a key factor in these transformations.[28] His argument that abortion was a sign of the rise of secular humanism and the collapse of Christian authority was transformative for many of the key figures in the Christian Right.

One of these men was Falwell, whose apocalyptic vision of American history shaped evangelical politics for two generations. Falwell was a different brand of religious leader than the Schaeffers. He began his ministry empire by founding a Baptist church in the 1950s. It soon grew to include Liberty University and a Christian television network. Through the 1960s Falwell was a leading evangelical segregationist, campaigning against the integration of public schools and directly criticizing civil rights activists, including Martin Luther King Jr. By the late 1970s Falwell had shifted his targets to the women's and gay rights movements. Drawing on a Calvinist tradition that viewed the United States as having a unique covenant with God, Falwell analyzed history, politics, and culture for signs that America had now fallen out of God's favor. In his 1980 sermon series *Listen America!*, Falwell called out the forces whom he saw as turning America away from its godly path. The legalization of abortion topped the list, followed by homosexuality, pornography, humanism, and the fractured family.[29] Through his political action committee, the Moral Majority, Falwell launched a legislative offensive that transformed Schaeffer's theological vision into a political juggernaut.

Antiabortion activism was at the heart of the Christian Right. Crucially, it was also integral to the institutional structures of the creation science movement. One of the founders of the Moral Majority, Timothy LaHaye, was also a founder of the most important creationism organization of the era, the Institute for Creation Research (ICR). LaHaye had been a Baptist minister in South Carolina and then Minnesota through the 1950s, and then, with so many of his generation, he had moved to the Sunbelt. In California he helped to found Christian Heritage College in San Diego and its sister organization, the ICR, in El Cajon, alongside Henry M. Morris. Morris, a civil engineer, was one of the two key founders, with the theologian John C. Whitcomb Jr., of young-earth creationism. In their 1961 book *The Genesis Flood: The Biblical Record and Its Scientific Implications,* Morris and Whitcomb argued that a literalist interpretation of Genesis could and should be supported by modern geological evidence.[30] The marriage between literalist biblical interpretations and this flood geology soon became known as creation science.

From its beginning in 1972, the ICR led the political agenda of creationism, supporting bills to mandate the teaching of creation science in public schools, publishing a creationist research journal, *Acts and Facts,* and providing churches with a Bible study curriculum to help their members articulate an antievolutionary apologetic.[31] The ICR was part of an evangelical campus in San Diego County, in Southern California, that included the four-year Christian Heritage College and the megachurch Shadow Mountain. These entities, in turn, had deep formal and informal ties to such conservative think tanks as the Heritage Foundation and other politically influential megachurches in the region. The ICR leadership, particularly LaHaye and his wife, Beverly, were key figures in the conservative campaigns against the Equal Rights Amendment and in pro-life activism during this period. Beverly LaHaye founded the Concerned Women for America in 1979 after watching an interview with Betty Friedan (an activist and author of the 1963 book *The Feminine Mystique*) by Barbara Walters on national television.[32] Incensed that the "humanist" morality of Friedan could take over America, through the Concerned Women for America LaHaye helped lead the most politically influential conservative women's movement in a generation.

Importantly, the moral perspective of the evangelical antiabortion movement was not associated with creationism merely by institutional proximity; it was one feature of a cohesive approach to national and global politics that saw secular science and progressive politics as twin threats to white Christian nationalism. Recognizing this situates evangelical antiabortion activism within the broader political worldview of white Christian nationalism. Thus, contests between evolution and creationism, segregation and civil rights, pro-choice and pro-life, LGBTQ rights and religious freedom, or immigrant rights and nativism are not isolated political matters. Rather, they are interconnected debates that have at their core the question of Christian authority in American culture.

Nevertheless, it took the *McLean* case for these politics to fully collide with the world of academic biology. In the years leading up to the trial, Morris and the ICR had worked diligently to introduce the teaching of creation science into public science classrooms. Morris's 1974 book *Scientific Creationism* was intended to persuade public school teachers that Darwinian evolution was faulty science and to serve as a manual for teaching creation science as a legitimate scientific alternative. Creation science efforts had succeeded in pushing a bill through the Arkansas state legislature that mandated "balanced treatment for creation-science and evolution-science" in Arkansas public school classrooms.[33] When the ACLU helped the plaintiffs, which included parents, religious organizations and leaders, teachers' groups and biologists, to bring a suit against the state, arguing that the bill violated the establishment clause of the First Amendment, the stage was set for the trial that brought Gould and other expert witnesses to the Arkansas State Supreme Court.

A New Scopes Trial

Even prior to *McLean*, there were many signs that the biological community had begun to take note of the actions of American creationists. In November 1981 the prominent American journal *Science* ran a piece by Roger Lewin titled "A Response to Creationism Evolves."[34] Beyond the pun in the title, the piece was meant to capture the changing response of the biological community to the creationist movement, as well as to spur further action. No longer would biologists be content to largely ignore what had seemed to be marginal publications and activities by creationists. Responding to the equal-time laws that would eventually prompt a reaction by the *McLean* plaintiffs, Lewin wrote a rallying call for action in light of the "bills already enacted in the states of Arkansas and Louisiana effectively mandating the teaching of the biblical account of creation" and declared that the "time is more than ripe for coordinated reaction by evolutionists and their supporters."[35] Lewin, who was an anthropologist, a scientific activist, and a collaborator of the notable evolutionist Richard E. Leakey, was the news editor for *Science* at the time.[36] Nor was this his first editorial on creationism. The previous year, he had published an article weaving together concerns over creationists and theoretical disagreements between various evolutionary camps.[37] The general tone of his pieces on creationism was one of unrest and disturbance.[38]

Within the evolutionary community, much of the immediate worry had been provoked by comments Reagan made during the 1980 presidential campaign. During a campaign stop in Dallas in August of the previous year, Reagan had addressed a gathering of ten thousand, many of whom were New Right preachers and educators. In his speech, he had claimed that the lack of religion in American public schools had contributed to human suffering, and he had declared evolution to be "a scientific theory only" that was "not believed in the scientific community to be as infallible as it once was." He had then added that if the theory of evolution was to be taught in public schools, so should the Biblical version of the origin of human life.[39] Reagan's subsequent electoral victory over Carter and the widely held belief that his triumph had been secured by the evangelical arm of the Moral Majority was a cause of serious concern for many evolutionists, particularly left-leaning ones. Reagan's speech created a ripple effect through the professional evolutionary community, especially those who were involved in popular writing, public speaking, or leadership of one of the community's professional societies. Gould, who fell into all three categories, was an active participant in these conversations. During the fall of 1980 and through 1981, in letters to various colleagues, Gould decried the fact that the country now had a "creationist in the White House."[40]

The rising concern over the political power of creationism was felt throughout the evolutionary biology community. In 1980, in the heat of an Arizona summer, the Society for the Study of Evolution (SSE) created an education committee to "study the current anti-evolutionary movement in the country."[41] Founded in 1946, the SSE was the primary professional organization of evolutionary biologists. The then president, Francisco Ayala, a population and evolutionary geneticist based at the University of California, Davis, put together the proposal for the committee.[42] He suggested John Moore, a biology professor at the University of California, Riverside, as its head. Moore, originally from West Virginia, was an active supporter of evolution-based biological education in public schools. He had been one of the founders of the Biological Studies Curriculum Study, which had promoted the teaching of evolution in public high schools.[43] Moreover, he had written numerous articles against creationism in such education periodicals as the *Journal of Science Teacher Education* and *American Biology Teacher*.[44]

Ayala, along with fifteen other members of the SSE, sent a letter to Gould inviting him to join the new Education Committee. In some respects, the arrangements for this anticreationist group followed a pattern that was typical for any SSE committee: there were discussions on members' tenure, the selection of the chair, and tasks for the group. The goals of the Education Committee, however, were unusual for the SSE. Instead of considering matters directly related to the practice of professional science, the committee's work was directed outward. Its members were tasked with controlling and disseminating information about evolution to the American public and to public classrooms. Ayala set out the committee's primary aims, which included gathering historical information on debates over evolution and preparing lists of experts willing to serve as expert witnesses in court cases and public debates.[45] Ayala, in other words, wanted an official arm of the professional society that was publicly active in the political struggle against antievolution activities. This was an ambitious set of tasks for evolutionary biologists, who were ostensibly occupied with all the matters that usually concerned professional researchers. But Ayala was confident that these matters would not take up too much of any individual's time. He envisioned a grassroots network of evolutionary sympathizers, with the professional biologists of the SSE serving as the experts at its helm, combating the very successfully organized creationist movement.

By the next summer, however, the committee had not yet gotten off the ground. The minutes from the 1981 meeting reveal frustration on the part of many in the organization's leadership that "little progress" had been made in the "six months since it was established."[46] Although there were individual scientists who were very much absorbed with anticreationist activities, organizing a professional society of biologists for a political aim was proving more difficult than anticipated.

Why did the SSE Education Committee meet with limited success? The SSE might have been expected to be the natural place for political agitation to take place. After all, the SSE had been the driving force behind the institutionalization of evolutionary biology in the years immediately following World War II.[47] For almost three decades the SSE had served as the center of professional organization and community building for the newly defined category of evolutionary biologists. But the SSE was not a tightly organized political advocacy group. For the first forty years of its existence it had had one primary aim: to further the professional and intellectual ambitions of evolutionary biologists. The organization could not summon either the moral will or the political capital to marshal expertise against a well-organized foe overnight.[48] Added to this, many individual members had little interest in public confrontations with the creationists.

Perhaps surprisingly, Gould himself was reluctant to engage with the creationists. He was concerned that debating them would simply legitimate their presence on the same public stage as evolutionary experts. In a letter from the summer of 1980 to Eric Christianson, at the University of Kentucky, Gould argued that it was "most unwise" for colleges and universities to organize debates between creationism and evolution "because creationists are so well organized politically to exploit this very situation."[49] Gould continued to be firm on this point. He repeated this theme in letters to colleagues around the country during this period, including a missive sent out to several colleagues in November 1981 in which he shared his "conviction" that it was "counter-productive to debate creationists face to face" for the simple reason that "one doesn't play the enemies' game."[50] Gould contended that the current creationist strategy was only the latest in a long line of failed attempts by political conservatives to circumvent the First Amendment and introduce religious fundamentalism in schools.[51]

An article by Lewin in *Science* in the fall of 1981, reporting from a meeting at the National Academy of Sciences, suggests that Gould was not alone in seeing debates with the creationists as counterproductive. He quoted Eldredge, who characterized what he felt to be the difference between the two communities as follows: "Scientists expect to have an exchange on rational grounds . . . but that's not how the creationists debate." Lewin also quoted Ernst Mayr, who charged that the creationists consistently misrepresented scientific facts: "They [the creationists] bring up the same old things again and again . . . such as the second law of thermodynamics and the bombardier beetle, which they must know now do not support their case. How do you counter this kind of thing?"[52]

The SSE did not succeed in organizing a decisive political advocacy group, but the plans for the SSE's Education Committee capture the growing sentiment within the evolutionary biology community that they needed to "do something." These fomenting concerns in the community gave legitimacy to the popular writing, magazine features, and television appearances of charismatic individuals such as Gould. Despite his initial hesitations, Gould quickly came around to the idea that creation science could not be ignored but must be defeated by direct attack. He was fighting the good fight for the community, and his popular activities in this moment were morally and professionally uncomplicated in a way unlike at any other time in his public career. When Gould went to Arkansas to testify against the creationists, he had the support of professional biologists around the country.

Gould began preparing for his testimony in the McLean trial shortly after returning from his trip to Dayton. In a letter written to several scientific colleagues, he explained that the subject of creationism was "all-consuming" but that he was nevertheless looking forward to joining "others of you in the Arkansas trial."[53] During his trial deposition, Gould wove his own experiences into a narrative that cast creationism as a long-standing evil in American public life. When asked whether he had studied evolution during his own days at a public high school in Queens, he commented that he had learned "virtually nothing" and that this was the "legacy of the Scopes trial and the textbooks which still exist not many years later." When asked to demonstrate how he had arrived "at the conclusion that [this] was a legacy of the Scopes trial," Gould replied, "Because the book we used was Moon, Mann, and Otto. And we know since that book was around since 1920 that, although early editions included much evolution, the post-Scopes trial ones did not."[54] Gould based his understanding of the sequence of these events on recent historical scholarship, though he gave it more authority by presenting it as his own experience.

After the trial, Gould continued to draw upon and popularize the work of professional historians in his assessment of the Scopes trial in his piece "Moon, Mann, and Otto."[55] The title was taken from the authors of the biology textbook that Gould had used as a high school student—an edition of the textbook from which evolution had been excised. He relied on the work of the historians and sociologists of science Judith V. Grabiner, Peter D. Miller, and Dorothy Nelkin (who was also a witness at the McLean trial) in drafting the piece.[56] Grabiner and Miller wrote one of the earliest histories linking the Scopes trial to changes in biology textbooks. Their article "Effects of the Scopes Trial," published in *Science* in 1974, documented changes to high school textbooks from editions published in the Progressive Era to those found in the post–World War II period. They concluded that the "impact of the Scopes trial on high school biology textbooks was enormous" and that "in the story of textbooks, the victory of the views of the scientific community is neither swift nor inevitable."[57] Their work, along with Dorothy Nelkin's book *Science Textbook Controversies and the Politics of Equal Time*, shaped Gould's understanding of the historical impact of *Scopes* and the place of *McLean* within that narrative.[58]

In his column "Moon, Mann, and Otto" Gould described himself as the "victim of Scopes's ghost" because he had attended high school in the era described by these historians. In his column, he portrayed himself as someone who had become a highly successful evolutionary scientist in spite of, rather than because of, *Scopes*. This personal quality of Gould's appeal had already been established by the tone of the earlier piece "A Visit to Dayton," which included photographs of him taken by his wife. Gould's determination not to let the aftermath of the McLean trial be like that of the Scopes trial was more than an academic or professional concern. Both in his writing and during the proceedings of the trial, Gould testified to the actual personal harm and lasting animosity engendered by the creationist movement's outflanking proponents of science and progressivism, on whose side Gould now counted himself. In his column, Gould equated contemporary creationists and evangelical Christians with William Jennings Bryan (the lawyer on the side of creationism in the Scopes trial): "Today Jerry Falwell has donned Bryan's mantle . . . of course we will not replay Scopes's drama in exactly the same way; we have advanced somewhere in fifty-six years. . . . But the similarities between 1925 and 1981 are more disconcerting than the differences are comforting."[59] By explicitly connecting the past and the present for his readers, Gould's historical narrative presented creationists as an intellectually stagnant but politically savvy group.

Although during the trial Gould was asked to comment on specific creation science theories, the key legal issue of the trial was the definition of *science*. In fact, the most discussed legacy of the McLean trial has been the philosophical principles that led Judge Overton to rule that creation science did not fit the characteristics of science but was religiously motivated. The case, in other words, helped develop a working legal definition for distinguishing science and religion in the American context. Overton's decision used pamphlets, subpoenaed internal memos, and textbooks written by the act's authors to argue that Act 590 conveyed an inescapable religiosity. But just as importantly, Overton also argued that creation science did not fit the philosophical criteria of science. For this point he relied heavily on the testimony of the professionally credentialed philosopher and historian of science Michael Ruse. Overton's decision defined sciences as guided by natural law, testable against the empirical world, having tentative conclusions, and being falsifiable. Overton determined that creation science failed to meet these criteria and therefore could not be science.[60]

From a legal standpoint, Ruse's testimony and Overton's decision were more important aspects of the McLean trial than Gould's participation.[61] But while these legal issues have had important consequences for American jurisprudence, the trial itself stands as a transformative moment in the relationship between science and politics in America. The publicity associated with the trial and the political activism of biologists in response to the trial turned the evolution-creation debate into a full-scale political matter.

After all, part of Overton's decision rested on the point that there was "not one recognized scientific journal which has published an article espousing the creation science theory."[62] It was crucial for the case to define a community of scientists as the keepers and definers of science. During the trial, science was defined as "what scientists do."[63] Mark Herlihy, one of the plaintiff lawyers, later argued that the expert testimony in *McLean* had been unusual from a legal standpoint. Herlihy explained that the "science and education experts" had not been "testifying about the absolute truth or falsity of any particular scientific fact." Rather, they had "presented to the court the results of twenty years of . . . claims of the creationists, set against one hundred years of experience since the Darwinian revolution."[64] Gould, as an established member of the specialized community of evolutionary biologists, was a crucial voice during and after the trial.

The McLean trial lasted two weeks, but Gould was so confident in victory for his side that he wrote an op-ed for the *New York Times* about the significance of the trial before the verdict was delivered. The piece was published three weeks later under the title "Evolution Wins Again," alongside two historical news items originally published in 1925, "Evolution, A 'Crime,'" by Clarence Darrow, and "Scopes: Infidel," by H. L. Mencken.[65] Darrow, of course, was on the side of evolution in the Scopes trial, and Mencken was a famed American journalist from the same period. Both articles expressed disbelief that creationism could gain a foothold in modern American society. Gould's editorial, paired with these voices from the past, helped to solidify his point that evolutionists should not allow history to repeat itself. Gould did not want evolutionists to make the same political mistakes as had befallen advocates for evolutionary theory after the Scopes trial.

In his editorial Gould also discussed what he considered the larger political implications of the current legal controversy. Positioning himself as the voice of the "professional evolutionists," he asked, "Why should anyone else view it with more than mild amusement?" Answering his own question, Gould argued that creationism "can only flourish in an ambiance of unquestioning authoritarianism." Implicit in this statement was his desire not to see American society travel down this road. He claimed that the "the growth of creationism" reflected "the successes of a larger political program (identified with the Moral Majority and other rightist groups) that include defeat of the Equal Rights Amendment and a total ban on abortion."[66]

It served Gould's purposes to cast creationism as a slumbering evil that was neither politically nor intellectually new; this helped him contend that creationism was neither a genuine scientific alternative nor representative of religion more broadly. But Gould's ongoing contention that there was nothing new about creationism mischaracterized its novelty, both as a theology and a political movement. The creationism of the 1980s was a powerful force that had deep institutional ties to the organizations of the religious Right; from these ties, it animated an increasingly visible public platform for the remaining decades of the twentieth century. And perhaps most significant for Gould's community, this moment inspired the first national anticreationist activist movement in the nation's history.

Anticreationism Activism Grows

The correspondence, publications, and trial preparations by Gould and other evolutionists in the early 1980s have an air of mobilization about them. History had crept up on evolutionary biologists, and they came together to do something about it, even if their professional society would not. The public debate with creationists made political involvement much less professionally or personally problematic for evolutionists than public involvement over such topics as sociobiology had been just a few years prior. The spectacle of the trial meant that part of being an evolutionary biologist was to be publicly anticreationist in a way that had been inconceivable a decade earlier. Although biologists had taken activist roles throughout the twentieth century, it was in this early 1980s moment that their professional duties included the defense of biology against religion. Overall, they accepted this call. Gould's participation in the trial and his anticreationist writings were part of a contest for authority over biology in public, political, and even legal forums.

Gould was one of the most articulate voices framing and conferring meaning on the events in Little Rock. Although he was certainly not the only person to compare the Scopes and McLean trials during the lead-up to and the aftermath of events in Arkansas, he was uniquely influential. After several years at *Natural History* and an increasing number of editorials for such mainstream press outlets as the *New York Times*, the *New York Review of Books*, and the *Boston Globe*, Gould was well connected in the world of press and publishing. He understood editors and editorial practices (despite his occasional protests to the contrary) better than most scientists in his professional circle did.[67] His connections with the editorial staff at *Natural History* and his ability to influence the crafting of a narrative around *McLean* demonstrate the importance of Gould's historical vision for understanding the unfolding events.

Gould's most famous piece from this period demonstrates his view that he was writing for a political conflict. His essay "Evolution as Fact and Theory," first published in *Discover* in May 1981, was later reprinted several times and continued to be heavily cited for the remainder of his career.[68] In this essay Gould contended that evolution was both an explanatory structure and a demonstrable empirical reality. He treated his readers to a brief lesson in the philosophy of science to make what he believed was an important distinction: "Well evolution *is* a theory. It is also a fact. And facts and theories are different things, not rungs in a hierarchy of increasing certainty. Facts are the world's data. Theories are structures of ideas that explain and interpret facts. Facts do not go away when scientists debate rival theories to explain them." By disabusing his readers of the notion that evolution was merely a theory (in the "American vernacular" in which "'theory' often means 'imperfect fact'"), Gould argued that evolution was not only a successful framework for understanding the mechanism for change in the history of life but also a well-established fact. In other words, although "evolutionists make no claim for perpetual truth," Gould asserted firmly that "no biologist has been led to doubt the fact that evolution occurred; we are debating *how* it happened." Gould wanted to persuade his readers that although science did not produce dogma, it could produce certain and reliable knowledge.[69]

"Evolution as Fact and Theory" was reprinted in the volume *Speak Out Against the New Right*, published by Beacon Press only a few short months after the McLean *trial* ended in 1982. Edited by Herbert Vetter, the lead Unitarian minister at the First Church in Cambridge, in Harvard Square, the volume contained twenty-two essays by a variety of liberal luminaries of the political and literary worlds, including Gloria Steinem, Carl Sagan, and John F. Kennedy Jr. The essays covered issues from birth control to environmentalism and heralded warnings that the evangelical Right was fast taking over American religion and politics. In his introduction to the volume, Vetter argued that the right-wing "counter-revolution is empowered . . . by the suave political electronic presence on stage of New Right President Ronald Reagan."[70] The introduction to the second edition listed the "New Right imperatives of power" that the forum believed conservatives advocated, which the forum existed to oppose:

Ban godless books from our libraries and schools
Get women back home and out of the job market
Make the Bible the foundation of our government and laws
Become the dominant power in the Congress, the courts, and the White House
Help fund religious schools through government vouchers
Oppose both contraception and abortion
Write creationism into science textbooks[71]

Although creationism was last, its appearance on a list of concerns about the "takeover" of the New Right is notable.

Speak Out Against the New Right grew out of a speaker series called the Cambridge Forum, held at the First Church in Cambridge from 1979 through 1982. The setting had a direct kinship with other left-leaning Cambridge circles in which Gould took part. The forum was broadcast over WGBH radio, the parent company of NOVA, to which Gould was closely connected. Richard Lewontin also spoke at the forums and was an even more enthusiastic participant than Gould.[72] Vetter, as a Unitarian minister, represented the progressive wing of American mainline religion, toward which Gould was much more sympathetic than he was toward the fundamentalism of the ICR. Gould was keen to separate the evolution-creation controversy from a general discussion of the relationship between science and religion and continued to see the debate with evangelicals as a narrow, political, sectarian division, not a true conflict of worldviews.

If *Speak Out Against the New Right* connected creationism to a New Right political agenda, Ashley Montagu's 1984 edited volume *Science and Creationism* directly used the McLean trial as an occasion to confront the creationist movement. Gould contributed "Evolution as Fact and Theory" as well as "Creationism: Genesis vs. Geology" (originally published in the *Atlantic Monthly)* to this collection. Gould had a close friendship with Montagu, and the older scientific activist often advised Gould and encouraged him in his progressive scientific activism. One of Montagu's strengths as an activist was his ability to collect and release edited volumes on controversial topics in a timely manner. This 1984 collection brought together essays from a variety of scientific figures, including the anthropologist Laurie R. Godfrey and the biochemist Sidney Fox. Issued by Oxford University Press, the collection was written by and for academics. It reprised much of the material that had been printed in the pages of *Science* directly after the McLean trial, including Judge Overton's decision and Roger Lewin's article. The message of the volume was clear: professional biologists were incensed at the claims and the political success of American creationists.

These pieces were just a few of the numerous publications, including popular books, academic volumes, conference presentations, and professional society newsletters, that Gould and others involved in anticreationist activities published in the years immediately after *McLean*. The trial became a rallying point for discussion, debate, and indignation for the professional community. Little of this, however, translated into concrete action. As Ayala had seen in his attempts to organize the SSE, it was difficult to get biologists to successfully commit to the work of political advocacy. Gould was an especially active public figure, but even his involvement and enthusiasm would need to be organized after the early 1980s. Eventually Gould and other scientists would channel their political activity through a new political organization founded by Stanley Weinberg, the National Center for Science Education (NCSE).

Stanley Weinberg was a high school biology teacher and science writer from Iowa who began organizing statewide Committees of Correspondence in response to the creationist movement in 1980. In his correspondence and research materials from this period, it is clear the McLean trial was a major motivation for the development of these groups.[73] Weinberg's organization went on to establish grassroots organizations in thirty-seven states by 1982, which ultimately incorporated to become the NCSE in 1983. Weinberg thrived at what the SSE had failed to accomplish: creating an organized system of professional evolutionists and science teachers to mobilize against creationist activities in local school settings and in court cases. Early on, Weinberg partnered with the ACLU, obtained grants, and recruited scientific luminaries for speaking engagements for the new organization. From its inception, the NCSE succeeded in reaching and rallying professional scientists.

In 1981, Roger Lewin's *Science* article "A Response to Creationism Evolves" introduced the Committees of Correspondence to professional scientists around the country. Lewin reported to the readers of *Science* that "sensing the need for grassroots action against legislative and other initiatives by the creationists, Stanley Weinberg . . . set up a network of committees of correspondence."[74] After this article ran, Weinberg received letters from many scientists around the country asking how to become involved in the new organization. In one such letter a Dr. John Mauldin of Iowa declared, "I would like to be in touch with you and your organization concerned with the creationists as a supporter."[75] Weinberg was keen to get professional scientists involved. An early memorandum circulated to the organization's leadership explicitly highlighted his desire for credentialed scientists. Under the memo item "recruitment," Weinberg suggested, "Scientists probably should form the nucleus of a C/C [Committee of Correspondence] because they have the expertise and the credibility that we need. But along with scientists, teachers and clergymen, do no neglect to recruit lay people of all types."[76] Weinberg hoped that experts such as Gould could partner with scientifically minded citizens around the country in the fight against creationists.

At the outset, Gould had only incidental contact with Weinberg and the Committees of Correspondence. Weinberg did send Gould a few letters in the years leading up to *McLean*, but the two did not have an extended correspondence.[77] During the trial itself, Gould was questioned about his participation in formal anticreationists activities. Typically, Gould was somewhat reluctant to identify himself as a representative of any of the newly formed anticreationist organizations. When asked at the McLean trial whether he was a participating member in the newly formed Committees of Correspondence, Gould replied, "In a loose sort of way yes. I have spoken very briefly to Stanley Weinberg in the State of Iowa, whom I understand is their leader . . . but have had no formal ties with them."[78] Gould preferred to present himself as an independent scientist and thinker before the court. However, his story intersects with the work of Weinberg and the NCSE in important ways. Weinberg and the NCSE were instrumental in rallying the events of *McLean* toward the cause of the new anticreationist movement. Although Gould wrote prolifically on creationism, it was the NCSE that created a stable, organized political advocacy network. Gould's continued political action against creationism was dependent on the political changes brought to his professional community, changes that were partly the work of the NCSE.

Although Gould was not heavily involved in the NCSE in its initial years, by the end of the 1980s he had developed a collaboration with the second director, Eugenie Scott. Scott, a physical anthropologist, took over the executive directorship in 1987. Her tenure saw a change in the organization's focus away from "creationism-bashing" toward wide-reaching education programs.[79] Scott explained the reasons behind this shift in a February 1988 letter to Gould, stating that the organization had originally been "crisis-oriented" but that as a result of the "slowing down of creationist efforts," the organization had moved to promoting public education on evolution.[80] By the end of the decade, Gould was increasingly involved in efforts for public education and understanding of evolutionary science, many of them directly connected to the NCSE. Although this 1988 letter from Scott captures a quiet political moment, there would be renewed bouts with creationists with the rise and promulgation of the intelligent design movement in the early 1990s.

During the McLean trial, Gould leveraged his professional status and his popular appeal to become a political champion for the evolutionary community. Gould thrived in this role. It was not that his testimony made much difference in the trial, nor even that his testimony was the deciding factor in the legal issues deliberated by the court. But Gould used the trial as a platform for promoting a vision of science, culture, and life as driven by freethinking progress. After his experiences with *McLean*, Gould remained committed to the notion that evolutionary theory belonged to the larger human ambition of advancement for science and society. A number of factors contributed to Gould's success and credibility as a political advocate in this moment: his credibility as a researching scientist, his charismatic appeal to a growing popular audience for his writing, and his ability to craft a historical narrative that put creation science in an intelligible context. But creationism's public appeal exceeded both Gould's star power and the NCSE's organizing might. For the rest of Gould's career, his own disagreements with his evolutionary colleagues would be bound up in the larger creationism controversy. This was a new era for evolutionary expertise in American culture and politics.

Scientific Populism vs. Scientific Expertise

During this period, Gould's public writing about the creation science movement combined dismissal and disgust. In "A Visit to Dayton," Gould described the supporters of creation science as a "motley collection" who used creationism to "reinstitute all the jingoism and distrust of learning that prepares a nation for demagoguery."[81] Gould's rhetoric prefigured a political tension that would echo through think pieces in the wake of Donald Trump's election more than thirty years later.[82] Did the evangelical right represent a disenfranchised working class responding to the oppression of a globalized capitalist elite? Or were they the most powerful entity in US politics, driven by patriarchal and white entitlement? Although Gould was not particularly interested in the racial dynamics of Reagan-era evangelicalism, he did tend to regard creationism as a unique form of southern Christian

American anti-intellectualism.[83] For his part, Richard Dawkins expressed his classist disdain for the movement, describing it simply as "redneck creationism" in *The Blind Watchmaker* (1986).[84]

The relationship between the white evangelical narrative of working-class disenfranchisement and the reality of its political power is a fraught issue, one that historians and political theorists continue to debate.[85] Although I will not offer a pronouncement on this point, I do suggest that we gain insight by analyzing creationism as a populist approach to scientific knowledge. As several histories have observed, during the early part of the twentieth century the American antievolution movement was explicitly connected to the populist politics of the era.[86] During the Scopes trial, for instance, William Jennings Bryan argued that school curricula should be decided by democratic majorities in communities rather than by elite experts or committees of academics.[87] The populism of the 1970s and 1980s creationist movement is less immediately obvious but becomes apparent once the movement is situated in terms of its relationship to a new era of evangelical partisan politics.[88]

We can think about the nature of the populist impulse in creationism by considering in more detail the kinds of institutions and activities that emerged out of the flood geology movement during the 1970s. When Morris went to San Diego in 1970 to help LaHaye set up the Institute for Creation Research, the institution's charter set out three main areas of focus: preparing teaching materials for elementary school children, researching the physical aspects of the flood, and conducting public outreach through radio programs, seminars, and literature.[89] In the first decade of the center's existence, however, its limited budget meant that its efforts focused primarily on disrupting the cultural authority of professional evolutionary science instead of conducting empirical research. This included, for instance, producing detailed reviews of the current scientific literature to scan for points that might be favorable to creationist arguments. The center was also a leader within the broader evangelical community in developing educational materials for children and young adults. Its publication *Acts and Facts,* which mirrored the tone and aesthetic of popular science magazines such as *Discover* and *Scientific American,* was widely circulated among evangelical churches, colleges, and other parachurch organizations. And the small library museum at the center's headquarters in El Cajon presented flood geology in a manner that would prefigure the establishment in 2007 of a much larger creationism museum in Kentucky, whose format and structure even more explicitly mirrored those of a modern natural history museum.[90]

Through these modes, creationists combined disruption and criticism of professional academic science with the self-creation of scientific knowledge and institutions. Strikingly, these efforts paralleled the actions taken by leftist activists in the same period. Consider, for instance, the publication of *Our Bodies, Ourselves* by the Boston Women's Health Book Collective in 1970 as an assertion of autonomy from male physicians by the women's health movement.[91] Or how the network of Janes in Chicago helped women gain access to—and later directly performed—safe and illegal abortions before the passage of *Roe v. Wade*. Or the Black Panther Party newsletter's publication of diagnoses of medical cases to fight against the structural racism experienced by Black patients in hospitals.[92] Or the Young Lords Party's active disruption of the administrative functions of hospital systems in Chicago and New York to provide better health care access for Puerto Rican communities in both cities.[93] In each of these cases, a populist impulse in the most basic sense—a grassroots movement of a community of people—acted against a form of expert science and medicine that the community viewed as oppressive. They made their own medical and scientific knowledge and educated their communities—about their bodies and the natural world—in the manner they saw fit. Even within the academic Left critique of science we can see parallels to creationist efforts in, for instance, the scouring of the professional scientific literature to scrutinize the evidence, models, and rhetoric of a field like sociobiology. Across the Left, communities used the same tactics—disruption, criticism, and self-creation—to protest the authority of professional science.

How do we make sense of the structural similarities between creationism and leftist health movements? Although there were exchanges between leftist groups (e.g., between the Young Lords and the Black Panther Party or among reproductive justice organizations), there is little evidence that creationists had institutional ties or direct exchanges with any of these organizations. Instead of assuming a sharing of political tactics, I suggest that we view the emergence of populist science on both sides of the political spectrum as a distinctive mode of claiming power within and against the state during a period of US history in which scientific expertise had become deeply intertwined with government authority during the Cold War.

In chapter 2, I referenced the basic historical realization that progressive movements mobilized scientific knowledge in the early twentieth century as part of the biopolitical transformation of the state in the modern world. The framework of biopolitics continues to offer critical scholars a vocabulary for the biological interventions that are at the heart of state action. Across such jurisprudential contexts as public health, immigration, reproduction, and environmental regulation, the state expresses its power through the scientific management of its population, determining who has access to life, representation, citizenship, and resources, as well as to bodily autonomy and privacy. Thus, it should not be surprising that communities across the political spectrum have turned to criticizing elite science or rejecting public health mandates as essential representative strategies. Engaging science and medicine, either through direct criticism or by creating grassroots alternatives, is essential to self-advocacy and claims to political power within the biopolitical reality of the modern nation-state.

And if we consider the rise of the military-industrial-academic complex of the United States during World War II and the Cold War, it becomes even more apparent why claims to political power in this era took the form of criticizing professional academic science. During these decades, science became indispensable to American nationalism and empire-building. The effort to research, build, and deploy the atomic bomb galvanized a new relationship between the federal government and American academic science. At the core of this was the Manhattan Project at Los Alamos—the largest scientific collaboration until that time, requiring more than 150,000 researchers and $2 billion—which built the weapons used in the bombing of Nagasaki and Hiroshima in August 1945. The Manhattan Project required an influx of funding and new models of cooperation between the academy, industry, and the federal government.[94] The war catalyzed a flood of defense funding into research labs across the country. The scale of funding and cooperation cemented science as the guarantor of American military superiority and its best defense against threats to American democracy.

By midcentury, technology fueled the nation's military dominance in a new global order. When World War II gave way to the Cold War, the threat of global communism solidified this new relationship between American science and American governance. As European colonialism was dismantled through the independence movements of the Americas, Africa, and Asia, the United States and the Soviet Union emerged as twin superpowers in a battle for economic and military supremacy around the globe. Cold War federal funding was initially concentrated in the sciences closest to defense interests, including nuclear physics, engineering, and weapons development.

During these decades Americans viewed science as a bulwark against communism, both at home and abroad. This manifested in two distinct ways. First, elites were concerned that poverty made domestic and foreign populations more susceptible to the influence of communism. This fear galvanized American efforts to track, manage, and solve problems of poverty. Policymakers and social scientists alike shared a view of malnutrition, overcrowding, access to water, birth rates, and education as properties of populations and as emergent demographic features that extended beyond any single individual. Particularly during the Kennedy administration, modernization theory shaped the creation of such programs as the Alliance for Progress with Latin America, the Peace Corps, and the Strategic Hamlet Program in Vietnam.[95] Heavily influenced by evolutionary models of social and biological progress, these programs framed Third World countries as "behind" in their social development. Certainly, developmental models of civilization had been influential in the United States and Europe since the late eighteenth century. But Cold War programs applied these frameworks in novel ways, through extensive consultation with social scientists, the creation of bureaucratic infrastructures, and the collection of social scientific data at an unprecedented scale.

The second battleground on which American science was to meet the forces of communism was much smaller; it was in the mind of the average American citizen. Elites looked to science to cultivate the intellectual habits and personal virtues that were necessary to sustain American democracy. It was a stunning transformation for a country that had long had confidence in religious belief as the moral basis for democracy. Since the founding period, American intellectuals had argued that a Protestant society was uniquely suited for democracy; Cold Warriors now contended that science was the surest way to cultivate a democratic citizenry. One such argument rested on the conviction that science engendered liberalism, that scientific practices helped instill an openness and flexibility of mind rather than the dogmatism and adherence to authority that American science advocates believed were the key features of communism. The universalizing arguments of liberalism—that American values were human values, and American rights were human rights—gave American elites confidence to expand America's reach on the global stage.

America's claims to technological supremacy were not solely an expression of military dominance; they were also a framing device for a specific version of a domestic social order. This was clear, for instance, during the famous "kitchen debate" between Vice President Richard Nixon and Soviet Premier Nikita Khrushchev. During the opening of the American National Exhibition in Moscow in 1959, the two leaders had a series of exchanges while viewing a re-creation of a suburban American home that proudly displayed a set of domestic appliances. Nixon argued that US domestic technology was a sign not only of the nation's technical prowess (in a clear metaphor for American nuclear capacity) but also of the prosperity of American capitalism, which allowed American housewives to remain at home. Conversely, Khrushchev asserted that the Soviet Union's superiority was demonstrated by its communist commitment to full gender participation in the labor force.[96] Indeed, one of the most enduring cultural touchstones for the American life in the latter half of the twentieth century was the image of suburban life that raised the baby boomers. This generation grew up in a world fashioned by postwar prosperity and technology and laced with the uncertainties of nuclear warfare. They built the infrastructure of suburbia, reinforcing racial divisions, and instilling a gendered vision of family life that would be idealized for the rest of the century. Science was at the heart of this vision, providing both epistemological grounding and the promise of technological security in the face of the ongoing threat from the Soviet Union.

In a context in which professional science had become so instrumental to state power, it is little wonder that both sides of the political spectrum made claims for political autonomy and community identity through scientific critique. But recognizing the structural similarity between the political tactics of creationists and those of leftist health movements did not mean that these groups had similar goals. Rather, it was *despite* their opposing social aims that they criticized science; it was through this tactic that these communities had their best hope of breaking free from the biopolitical state. But these were very different movements, with different constituencies, different political goals, and different alternatives to the authority of science. When the Young Lords and the Black Panther Party protested institutionalized medicine and science, they did so as part of the work of Black and Brown solidarity, survival, and self-determination, whereas creationists attacked Darwinism as part of the nostalgic apocalypticism of the evangelical Right, a cultural effort that combined a desire to return to a patriarchal and ethnonationalist American past, even while preparing for the end of days.

I believe our response to scientific populism must take into consideration the wider political context of any given populist movement. That is, it is not enough to ask whether individuals or communities ought to have the right to question professional scientific authority in the abstract. Indeed, because scientific authority has become so intertwined with state power, protesting the one is bound up with protesting the other. It is for this reason that this history has continually considered the vision of America that was bound together with leftist, liberal, and right-wing approaches to science. These communities understood the nature and purpose of science differently, and they articulated conflicting views on its relationship to the ethical project of American democracy. Creation scientists and intelligent design proponents believed that the materialism of evolutionary science encouraged moral relativism, which they blamed for contemporary vices in American culture. Undermining evolution was a critical part of their campaign to return America to traditional "family values." Liberal public intellectuals argued that science was the necessary foundation of democracy. It had been through Enlightenment rationality that humanity pulled itself out of a mire of superstition and ignorance into the light of modernity. Science was humanity's best pathway forward and ought to be defended against cultural doubt in all its forms. Leftist scholars and activists contended that science was complicit in the worst sins of our nation's history. They saw the hand of scientific racism in slavery and the work of eugenics in the poverty and disenfranchisement of generations of immigrants. These evils were so significant that it was incumbent on experts to expose the limits and faults of science to a wider public. When debating empirical matters about evolution—the age of the earth, the provenance of natural selection, or the evidence of the fossil record—these groups were each committing to a vision of American democracy as well as building an understanding of the nature of science.

The early 1980s were a critical pivot point for the evolutionary biology community. The events of the McLean trial both shocked and galvanized these scientists into seeing a new moral duty to be publicly anticreationist. This had the simultaneous effect of uniting liberal and leftist biologists in their public efforts against creationists, even as it raised the stakes for technical debates over the future of evolutionary theory. Across the scientific community, the ongoing public presence of creationism, along with the political power of the New Right, would recalibrate the terms of debate from the years of the sociobiology controversy. The conflicts between leftists and liberals did not abate; instead, they were exacerbated and recast once the role of religion came front and center in public debates over the authority of science in American culture.

CHAPTER FIVE

The Accidental Creationists

> I sense that some now wish to mute the healthy debate about theory that has brought new life to evolutionary biology. It provides grist for creationist mills, they say, even if only by distortion.
>
> —*"Evolution as Fact and Theory," 1981*

In the late summer of 1999 the Kansas State School Board voted to remove evolution, along with the big bang theory, from its statewide science requirements.[1] Although this action did not technically forbid the teaching of evolution, it was meant to effectively end it by excluding the scientific theory from what students were required to learn in public science classrooms. Writing for *Time* a few weeks later, Gould ridiculed the decision as a "misguided effort" that "saddens both scientists and most theologians."[2] As he had done during the early years of his battle with creation science, Gould characterized the moment as nothing more than the political work of a fundamentalist religious group. The Kansas State School Board decision had not been directly orchestrated by the Discovery Institute, the newest player in the network of creationist think tanks; rather it came about as a result of the broader intelligent design (ID) movement, which had been gaining steam since the late 1980s.[3]

ID advocates argued that the pervasive naturalism in modern science ought to be challenged and, moreover, that looking for evidence of supernatural design should be an acceptable practice in professional biology. In popular books such as Phillip Johnson's *Darwin on Trial (1991)* and Michael Behe's *Darwin's Black Box (1996)*, ID proponents argued that Darwinian natural selection was inadequate to explain the complexity of living things.[4] And by eschewing overtly Christian language, the movement was able to secure a series of temporary triumphs (including the Kansas decision) until the landmark Pennsylvania District Court case *Kitzmiller v. Dover Area School District* ruled in 2005 that ID was in fact motivated by religion and therefore could not be taught as an alternative to evolutionary science in public schools.[5]

Gould's writing about the Kansas state curriculum for *Time* was typical of the balance he struck during this period in his public defense of evolution against creationism. He was careful to insist that neither creation science nor ID represented any real threat to the validity of Darwinian science. He depicted the movements as the cynical work of conservative religious groups, while also clearly stating his "abiding respect for religious traditions." And he argued (incorrectly, as it turns out) that creationism was a uniquely American foible and not a political issue that other Western countries dealt with.[6] In all this, Gould cultivated a platform for himself whereby he could defend evolution, while also leaving space to advocate for his own scientific and political criticisms of modern Darwinian science. Gould, however, was not the only public intellectual who editorialized the moment. The Kansas State School Board's decision prompted another high-profile piece, published in December of that year in the *New Yorker*. But that piece accused Gould himself of being "bad for evolution" and "aiding and abetting" the creationist cause.[7]

Written by the journalist and public science writer Robert Wright, the essay was titled, "The Accidental Creationist." Wright acknowledged that his claim that Gould had "lent real strength to the creationist movement" would likely surprise many of his readers, since most of the nation regarded Gould as, in Wright's words, "America's unofficial evolutionist laureate."[8] What, then, was the basis for Wright's self-consciously audacious attack on Gould? Wright zeroed in on Gould's campaign against the centrality of natural selection to evolutionary explanations. As discussed in previous chapters, Gould's critique of adaptationism and his promotion of alternative evolutionary models shaped much of his technical and popular writing in the two decades after the publication of "The Spandrels of San Marco and the Panglossian Paradigm" in 1979. Gould directed these models at the logic of sociobiology, but during this period they were seized upon by creationists (as well as by journalists) as evidence that there were cracks in the foundation of Darwinism.

By 1999 Wright was frustrated enough by this dynamic to argue that Gould was himself at least partially to blame for creationism's political strength. But he was not the first public intellectual to complain about Gould's scientific self-advocacy. As early as 1986 Dawkins had devoted an entire chapter of his book *The Blind Watchmaker* to his contention that Gould and Eldredge's theory of punctuated equilibria was not a real departure from neo-Darwinian orthodoxy.[9] In 1986

Dawkins gave Gould and Eldredge some degree of grace; he acknowledged that they were both "doughty champions in the fight against redneck creationism" and sympathized with the experience of having one's nuanced scientific arguments mischaracterized by the journalistic impulses of the public sphere.[10] But in 1995 the philosopher Daniel Dennett expressed far less patience with Gould. In his book *Darwin's Dangerous Idea: Evolution and the Meaning of Life,* Dennett argued that Gould's theories did not represent a real challenge to the truth of Darwinian orthodoxy.[11] He claimed that Gould was not, in fact, a scientific revolutionary but only a thinker who was unable to accept the full implications of "Darwin's dangerous idea."

Creation science and its successor, ID, moved Darwinists to interpret their debates over evolutionary mechanisms not as the result of alternative research programs or differences in disciplinary training but as the hill on which scientific naturalism was to live or die. In this new political climate, Gould's criticisms of adaptive reasoning were consistently interpreted by both creationists and other evolutionary celebrities as attacks on the ability of natural selection to explain the world without supernatural design. Just as creationism had rallied the political will of evolutionists in remarkably short order, the consciousness of creationism rapidly changed the tone of their technical deliberations.

Creationism brought religion as well as the increasingly powerful Christian Right to the forefront of American debates over scientific authority. This combination exacerbated the earlier tensions between leftists and liberals. Even as leftists continued to translate the social commitments of political activism into critical examinations of science, technology, and capitalism, scientific liberals began to characterize these moves as being aligned with creationism.[12] Despite the lack of political similarity between creationists and most academic feminists, liberals began to characterize both groups as antiscience, antirationalism, and anti-Enlightenment. Once the public defense of science was cast in liberal terms, liberals increasingly depicted leftist approaches to science as bad-faith attacks on the truth of science. This was especially true of feminist critiques of evolutionary explanations of sexuality and sex differences. Although the feminist approaches to biology had originated in the 1970s, largely on the part of researchers within the natural sciences, by the end of the century some liberal science advocates would dismiss these critiques as both ignorant and dangerous.

Were Gould and other leftist academics (at least partially) responsible for the political success of American creationism? Does the critique of scientific racism and sexism in fact provide fuel to reactionary political attacks on professional science? And, if so, what social harm do these attacks represent? At the beginning of the twenty-first century the conflicts over scientific authority among leftists, liberals, and right-wing religious groups had their most tangible effects in debates over the teaching of evolution in public schools. But two decades later these political divides continue to shape policies that directly affect our daily lives, particularly in relation to reproductive justice, vaccinations, climate change, and environmental regulation. By reflecting on this moment when Gould and others on the academic left became "accidental creationists," we can consider whether the leftist criticism of science damages the public authority of science.

Punctuated Equilibria and Creationism

After the McLean trial, Gould's theory of punctuated equilibria emerged at the center of a professional conflict among evolutionary biologists, even as it was simultaneously implicated in the public conflict between those scientists and creation science advocates. Gould and Eldredge had first advanced their theory in an essay in Thomas Schopf's *Models in Paleobiology* in 1972.[13] Their theory explained the gaps in the fossil record as positive evidence rather than missing information. These rereadings of fossil evidence helped paleontology claim a new importance in evolutionary theory, but they also closely resembled creationist criticism that paleontological data could not definitively prove evolution. In the 1985 edition of *Scientific Creationism*, Henry Morris claimed that the "leading evolutionist" Stephen Jay Gould had proposed a new theory that supported what creationists had "long argued": that there were "no true transitional forms in the fossil record."[14] And the resemblance between the criticisms of creation science advocates and the actual structure of Gould and Eldredge's theory caught the notice of the American media in the flurry of journalistic attention in the wake of the McLean trial.

When Gould and Eldredge introduced the scientific community to punctuated equilibria in 1972, it received little recognition outside professional scientific circles. It was five years later that the wider public was first introduced to punctuated equilibria, when Gould dedicated a *Natural History* column titled "Evolution's Erratic Pace" to discussing the theory.[15] Notably, in this early description of the theory, Gould implied to his readers that it only modified the rate of Darwin's evolutionary model. He began his story on November 23, 1859, when, "a day before his revolutionary book hit the stands," Darwin received "an extraordinary letter" from his friend and fellow evolutionist Thomas Henry Huxley.[16] Huxley hailed Darwin for his theoretical achievement, but he was concerned, and Gould concurred, that Darwin had unnecessarily loaded his new theory down with the assumption that evolution happened gradually. "Natural selection," Gould declared to his readers, "required no postulate about rates; it could operate just as well if evolution proceeded at a rapid pace."[17]

This appeared to be a low-stakes matter. After all, why would it matter if evolution happened slowly or quickly? But there was more to Huxley and Darwin's discussion than the pace of evolution. In a part of Huxley's letter that Gould did not quote, Huxley informed Darwin that in contrast to Darwin's belief that "Nature makes no leaps," he believed that nature did, in fact, "make *small* jumps."[18] In other words, Huxley and Darwin were discussing the character of evolutionary change and a crucial aspect of the persuasiveness of Darwin's overall theory. The success of Darwin's argument against divine agency depended on persuading his readers that large-scale speciation in life's history was the result of an accumulation of smaller, observable changes within organisms. From his correspondence with Huxley, it appeared that Darwin wished to avoid any suggestion that his theory required leaps from one form to the next. Gould's new theory could be interpreted as reintroducing these jumps into the picture of evolutionary change.

But in "Evolution's Erratic Pace" Gould tried to persuade his readers that gradualism had never been necessary for evolutionary theory. By casting this argument in the words of the renowned evolutionist Huxley, Gould created the impression that other evolutionists had also objected to Darwin's gradualism in good conscience. In the rest of the article, Gould argued that Darwin's adherence to gradualism was the result of Western cultural preference and not an empirical conclusion. According to Gould, this view had negatively affected generations of paleontologists, who had "paid an exorbitant price for Darwin's argument." They had been forced to treat their data as so incomplete as to be nearly useless, at least where evolution was concerned. But Gould offered a "resolution" to "this uncomfortable paradox." "The modern theory of evolution," Gould wrote, "does not require gradual change . . . it is gradualism that we must reject, not Darwinism."[19] In this column, Gould made an argument that he would reiterate for the next two decades, that the theory of punctuated equilibria was a necessary correction that preserved the essence of Darwin's theoretical innovation.

A few years later, however, punctuated equilibria burst onto a wider media landscape. In 1980 an influential group of paleontologists gathered at Chicago's Field Museum of Natural History for a conference on macroevolutionary theory in which the theory was a main topic of discussion.[20] The meeting was widely covered by both the scientific and the national press, with reporters present from *Science, Newsweek,* and the *New York Times.*[21] But to Gould's everlasting fury, some of these reporters implied that his and Eldredge's work lent strength to prominent criticisms of Darwinism by creation science advocates. In Gould's retrospective account of the meeting, he blamed journalists for "grossly misread[ing] the . . . meeting as a sign of deep trouble in the evolutionary sciences" and therefore "an indication that creationism might . . . represent a genuine alternative."[22] For Gould, this was the moment when punctuated equilibria not only became associated with "the death of Darwinism" but also became "the public symbol and stalking horse for all debate within evolutionary theory." And since "popular impressions now falsely linked the supposed 'trouble' within evolutionary theory to the rise of creationism, some intemperate colleagues began to blame Eldredge and me for the growing strength of creationism!" At the hands of the journalists, Gould and Eldredge, along with their theory, were maligned as the primary assassins of Darwinian evolution.[23]

Or so Gould implied in this passage from *The Structure of Evolutionary Theory* (2002), written twenty years after the fact. Although he did not entirely blame one source for what he felt was a gross misinterpretation of the state of scientific debate in evolutionary biology, he did single out an article by James Gorman, "The Tortoise and the Hare," which ran in *Discover* just before the Chicago meeting.[24] Outraged at Gorman's claim that fundamentalists were using disagreements among scientists to justify creationism, Gould and Eldredge published a letter to *Discover*'s editor "vigorously" protesting "this misrepresentation of our views."[25]

Both in 1980 and in *The Structure of Evolutionary Theory*, Gould portrayed himself as the offended party. What is notable about this incident is that Gould himself was involved in the writing of "The Tortoise and the Hare." From 1979 until 1982 Gould corresponded with the writing and editorial staff of *Discover*. The magazine itself was brand new, having released its first issue to stands in October 1980. A subsidiary of *Time* magazine, it was proposed and promoted by Leon Jaroff, who had worked as an editor for the larger news magazine. Jaroff envisioned *Discover* as a science news magazine, an idea that he developed once he noticed that the newsstand sales of *Time* magazine shot up when the cover story was about science or medicine.[26] Jaroff cultivated a relationship with Gould, writing to the scientist in 1978 that "we will be delighted to print almost any article your fertile mind conceives."[27]

In the first fourteen months of the magazine's existence, it printed seven features on the evolution-creation debate, all written either by Gorman or by Gould himself. Gorman was a staff writer for *Discover*; originally a freelance writer, he had made his first journalistic break with a review of Gould's first two books (*Ontogeny and Phylogeny* and *Ever Since Darwin*) for the *New York Times* in 1977.[28] Gorman did consult with Gould during the writing of "The Tortoise and the Hare," but he felt editorial pressure to emphasize the potential role for creationism in Gould's debate with other Darwinists.[29] Later, Gorman came to regret how strongly he had framed the article, but he still felt that Gould had had significant input in the publication's take on the scientific situation.[30] And just a year later, *Discover* named Gould its first scientist of the year, complete with a profile written by Gorman.[31]

Gould, in other words, actively worked to ensure media coverage for punctuated equilibria, and much of this coverage portrayed Gould as an iconoclast. In May 1982 *Newsweek* featured Gould in a cover story, "Mysteries of Evolution," that presented Gould as a young, exciting revolutionary in the world of science. The author, Jerry Adler, worked with Gould on the article and believed at the time and afterwards that he had adequately captured Gould's place in evolutionary biology. This article was more careful to distinguish Gould's criticism of Darwinian orthodoxy from those of creationists, but it still emphasized the potentially groundbreaking nature of Gould and Eldredge's theory of punctuated equilibria: "It is an embarrassment to Gould when he occasionally finds himself enlisted, without his consent, in the creationist attack on evolution." In the wake of *McLean*, stories about Darwin were selling magazines, and Gould made an exciting and interesting subject for a profile.[32] According to Adler, the editorial staff at one point considered using the title "A New Theory of Evolution" but ultimately decided that it was too controversial. This piece, like many others, promoted Gould as a scientific celebrity on the basis of his criticism of Darwinian orthodoxy.

Despite his connections in the world of scientific publishing, Gould could not control the public life of punctuated equilibria. Beginning in 1980, writers from many different disciplines and occupations invoked punctuated equilibria in a variety of genres and media forms, for varied goals. Creation scientists cited punctuated equilibria in tracts for their grassroots campaigns at churches as proof of dissent over the veracity of evolution among scientists.[33] ID proponents published popular books that cited Gould's theory, as well as those of his critics, in support of their intellectual agendas.[34] Gould's scientific critics, most notably Dawkins and Dennett, published popular books that dismissed punctuated equilibria as nothing more than a distorted and overhyped version of orthodox Darwinism. Philosophers of biology, including Kim Sterelny and Elliot Sober, worked through the theory to identify essential points of difference between Gould's and other formulations of Darwinism.[35] And both journalists and professional historians chronicled the exchanges between Gould and his critics, dubbing the vitriolic debate the "Darwin Wars."[36]

These disparate conversations were all connected by their concern over the status of the Darwinian revolution. Both scientific veracity and the place of scientific debate in relationship to the American public were at stake. The very existence of the conflict between Gould and his evolutionary opponents destabilized the authority of biology. If biologists could not agree on evolution, how could the rest of the country be certain that it was true?

A Higher Darwinism

As the McLean trial ended in the early months of 1982, it featured on the cover of *Discover* magazine. Overlying a portrait of Charles Darwin, the headline read, "Darwin on Trial: Creationism vs. Evolution in Little Rock."[37] Darwin looked both serious and plaintive; holding the hand of a small ape on the scales of justice, he appeared to be pleading with the magazine reader on behalf of evolution. This tableau, alongside the phrase *creationism vs. evolution*, is by now so familiar that it is difficult to believe that the combination was new in the 1980s. Although caricatures of Darwin had been widespread in late Victorian England, in the United States evolution was typically depicted through imagery that conveyed various cultural preoccupations.[38] It was not until after the McLean trial that Darwin's image featured more frequently in twentieth-century American popular culture. The trial coincided with the one-hundred-year anniversary of Darwin's death, and evolutionary biologists, historians, and popular science writers seized the opportunity to celebrate the Victorian naturalist.

The celebrations of Darwin in 1982 entangled the creationist threat with scientific debates over the state of the evolutionary synthesis. Both in Gould's correspondence and in the materials published from various centennial events, it was clear that the controversy with creationists was fresh in the minds of those keen to promote the cause of evolution in the American context. In a letter to Paul Kurtz, a professor of philosophy at the State University of New York at Buffalo, Gould proclaimed that he was "well aware" of the upcoming anniversary, "especially after a fascinating day spent testifying in Arkansas last week." Although Gould was confident that he and the other evolutionists "had routed the creationists," he also worried that the trial would "not end the issue," and so he "applaud[ed]" Kurtz "on the topic of your Darwin centennial meeting."[39] The meeting in question was sponsored by the Council for Democratic and Secular Humanism and focused on the triumph of secular science over the creationist worldview. Kurtz was well known in philosophical circles for his development and advocacy of secular humanism, which asserted the ethical capacity of humanity in the absence of supernatural beliefs.[40] These centennial events connected historical and scientific reflections on the importance of Charles Darwin and his theories to the authority of science in contemporary American society.

This was not the first time that evolutionists had used a significant anniversary to stake a claim for Darwin's continued relevance. The first Darwinian centennial celebrations, in 1909 (commemorating Darwin's birth, not his death), had masked the "state of disarray" that accompanied the challenges to Darwinism from the newly christened field of genetics.[41] With the rediscovery of Gregor Mendel's work in 1900, biologists had worked to put together an empirical science that would explain the laws of heredity. For a time, it appeared that the power of genetic explanations might cause natural selection to fall out of favor in evolutionary science. By the 1930s, however, a group of biologists had begun to find a way to reconcile Mendelian genetics with natural selection via population genetics.[42] The publications of Theodosius Dobzhansky, Ernst Mayr, George Gaylord Simpson, and Ledyard Stebbins launched what became known as the evolutionary synthesis by translating mathematical models in population genetics into a workable research program. Together, these biologists defined the central dogma of the modern synthesis, that natural selection was the primary force of evolutionary change. By the 1950s Dobzhanksy's famous assertion that "nothing in biology makes sense except in the light of evolution" was the driving principle of the new evolutionary biology.[43]

The synthesis was more than a theoretical proposal; it was also an institutional force. Led largely by Mayr from his post at Harvard University, the new evolutionary synthesis influenced university departments, organized a professional society, utilized federal grants, and founded the journal *Evolution*.[44] The architects of the synthesis turned as well to the power of history. Beginning in earnest with the 1959 centenary of Darwin's *On the Origin of Species*, Mayr helped lead a program of retrospection on the history of Darwinism.[45] More than an effort to be clear about the past, Mayr's historical work demonstrated to future biologists that the synthetic program was complete and triumphant. The 1959 celebrations reinvented Darwin as the father of a unified evolutionary synthesis, proclaiming that the science of theoretical population genetics was entirely compatible with the mechanism of Darwinian natural selection; indeed, that selection was triumphant. Mayr continued to convene conferences and wrote several books promoting this narrative into the 1970s and 1980s.[46]

When Gould joined Mayr's department at Harvard in 1967, he encountered these historical reflections on the achievements of the evolutionary synthesis.[47] But Mayr's claims to the contrary, the synthesis had not been truly reconciled.

Gould entered the first decade of his professorship at a time when many fields were struggling to incorporate new ways of organizing, analyzing, and visualizing biological evidence into the Darwinian vision. Although Gould did not consider sociobiology to be a challenge to the evolutionary synthesis, Wilson viewed his own work as a reproof of the lack of behavioral studies in the synthesis.[48] Wilson even described himself as having been "radicalized" in the 1970s by a desire to push beyond the evolutionary synthesis into new territories for evolutionary research.[49] Other research fields more deliberately minimized the role of natural selection in evolution. New techniques in molecular biology raised questions about current understandings of the genetics of speciation, as well as the shape of the tree of life.[50] New journals in paleontology, embryology, and microbiology asserted new theoretical visions and established new angles of biological inquiry. Many other biologists in Gould's generation expressed their dissatisfaction with the synthesis by creating new institutional and intellectual spaces for novel research programs.

Nevertheless, when Gould began framing punctuated equilibria as a critique of the synthesis around 1980s, he brought an unmatched rhetorical skill and comprehensiveness to his accusations. He simultaneously declared that the synthetic Darwinian vision was "dead" and that Darwin himself had to be revised.[51] It is conceivable that Gould's career would have developed differently if he had tempered his criticisms of the synthesis or if his celebrity had not amplified the consequences of his pronouncements. But Gould depicted scientific progress as happening through dramatic shifts in worldview. He was convinced that scientific paradigms could be altered through the active acknowledgment of cultural bias, including scientists' historical narratives. If a higher Darwinism were to be realized, it would first be necessary to take down its existing theoretical structure.

Even as Gould worked to promote himself as a reformer of Darwin, he helped to fashion a popular market for Darwin and modeled devotion to Darwin as intrinsic to the professional identity of an evolutionary scientist.[52] In his public writing, Darwin emerged both as Gould's hero in the battle against racism and as a figure in a cautionary tale that cultural context affected all scientists, even the father of evolutionary biology. Gould suggested that present-day scholars could free Darwinism from these cultural biases by carefully studying Darwin's social milieu. Here Gould echoed the same principle that he was simultaneously developing in *The Mismeasure of Man* regarding racial bias in science. Social context inevitability influenced the formation of scientific theories; however, scientists could advance science by carefully examining social prejudices either in themselves or in the history of science.[53] Gould meant to complete Darwin's revolution by extracting natural selection from the mire of an earlier century's cultural prejudices regarding progress.

Gould was convinced that Darwin's evolutionary theory, interpreted correctly, removed the possibility of enshrining humanity as the pinnacle of evolution or ranking races, genders, and civilizations on a scale of universal worth. Gould contended that natural selection had eradicated progress because the theory only proposed "*local* adaptation to changing environments."[54] He argued that natural selection contained "no perfecting principles, no guarantee of general improvement" and that Darwin's view of "improved design" contradicted "the cosmic sense that contemporary Britain favored."[55] Yet Darwin had so clearly benefitted from Victorian social hierarchies, Gould argued, that Darwin himself could not acknowledge the full radical potential of natural selection. It was the "unresolved inconsistency" between Darwin the "intellectual radical" and Darwin "the social conservative."[56] As a "patrician Englishman at the very apex" of "Victorian Britain at the height of industrial and colonial expansion," how could Darwin "abjure the principle that embodied this triumph"? Gould portrayed the hints of progress in Darwin's work as lingering class biases in his psyche that explicitly contradicted "the bare bones mechanism" of natural selection.[57]

Through his exegesis of nineteenth-century evolutionary theory, Gould participated in new textual engagements with Darwin's oeuvre that brought another distinctive element to the 1982 centennial. One of the noteworthy features of earlier centennial celebrations had been the complete absence of historians (and, by extension, the relatively few biographical considerations of Darwin).[58] In the intervening period, a wealth of new manuscript material had helped fuel a surge of historical interest in the life and science of Darwin. In his popular writing, Gould drew on this manuscript material, as well as on scholarship by professional historians of science who sought a closer understanding of Darwin's theory through an intimate, even minute acquaintance with Darwin's life and letters.[59] Gould invited his readers to become acquainted with Darwin through history rather than solely through the power or influence of Darwin's theory of natural selection. He added his voice and celebrity to the image of Darwin developing in historical research, that of a Darwin who could be known on a personal level and appreciated for his complexity and contradictions.[60] But Gould was not simply an interested scientist who kept tabs on the field of Darwin studies. To the extent that Gould was involved in the historical recovering and reconsidering of Charles Darwin, it was in the service of his own view of how Darwinism ought to be restructured.[61]

Gould's attempt to remake the Darwinian revolution echoed his Kuhnian understanding of the nature of science, which also animated his efforts in *The Mismeasure of Man*. In his study of racism and nineteenth-century intelligence testing, Gould had claimed that scientists could prune their science of social bias by reflecting on their own prejudices. But Gould's ambitions went beyond socially conscious science. In the conclusion to *The Mismeasure of Man*, Gould argued that "debunking" could be a "positive science." In the vein of a Kuhnian paradigm, Gould contended that "science advances primarily by replacement, not addition."[62] In this view, it was incumbent upon scientists to constantly re-examine and critique older scientific ideas, not simply to accumulate new empirical facts to fit into existing theories but to generate new ones. In other words, Gould's evolutionary heresies were not meant to return contemporary evolutionary biology to a pure or originalist reading of Darwin's theories, nor did he intend to only map out new avenues of empirical research. When he expounded on the necessity for a hierarchal theory of evolution, he believed that this was a grander, higher vision of evolution that Darwin's original theories had not captured.

The Midwife of Natural Selection

The growing publicity surrounding Gould's version of Darwinism and its potential connection to creationism altered the landscape of popular accounts of evolution during the 1980s and 1990s. The prominence of Gould's views meant that other evolutionists now had the dual task of defending evolution against creationism and reassuring the public that Gould's theoretical revisions did not undermine the central core of Darwinism.

Gould described natural selection as he did all scientific theories: as literary propositions about the natural world that ought to be carefully read and revised by new generations. To Gould, the rhetorical technique, historical language, and period imagery Darwin had employed were all part of the theory itself.[63] Other evolutionists, by contrast, imagined Darwin's theory as an algorithm that, once discovered, unlocked the secrets of biology and the meaning of life. Two of the most notable figures who promoted this idea were Dawkins and Dennett. In his 1995 book *Darwin's Dangerous Idea* Dennett described Darwin's theory as a "universal acid" that ate "through just about every traditional concept, and leaves in its wake a revolutionized worldview."[64] Darwin's historical context was irrelevant to the importance of his idea. Dennett argued, "The actual historical man does fascinate me . . . but he is, in a way, beside the point. He had the good fortune to be the midwife for an idea that has a life of its own."[65] For Dennett as for Dawkins, natural selection was not a work of prose to be poured over and analyzed. It was a reality about the world that could only be acknowledged or denied.

By the time of the McLean trial, Dawkins had already made a name for himself in evolutionary circles with the publication *The Selfish Gene* (1976), a "full-throated" argument for the gene-centered account of evolution. But in neither *The Selfish Gene* nor its follow-up, *The Extended Phenotype* (1982), did Dawkins expend much energy arguing either against creationists or against Gould. Before *McLean*, Dawkins's responses to Gould's critiques of adaptationism had been conciliatory or at least tolerant. In *The Extended Phenotype*, for instance, Dawkins had suggested that different evolutionary fields might differ simply in emphasis and interest, with "molecular geneticist[s] interested in gene substitutions" and "paleontologist[s] interested in major trends."[66] Dawkins stressed that "there is indeed, much more agreement than the polemical tone of recent critiques would suggest." He dismissed the "few genuine opponents of Darwinism" and proclaimed that "we are all in this together"—those "who substantially agree on how we interpret what is, after all, the only workable theory we have to explain the organize complexity of life."[67] The phrase "few genuine opponents of Darwinism" was the only allusion to antievolutionism in the book; these enemies were inconsequential enough that there was space for professional differences in biology. His adaptive logic was not yet primarily aimed at a religious opponent.

But the events of this period dramatically reoriented Dawkins's intellectual arguments about Darwinian evolution; they likewise set the stage for his later role as a leader of the New Atheism movement after the events of September 11. In 1986, with *The Blind Watchmaker*, Dawkins turned the intellectual task of Darwinian evolution into a remarkably focused mission: to demonstrate the insufficiency of the theistic argument from design. In the book, Dawkins presented the appearance of design in the living world as the most important mystery that had ever confronted humanity. Throughout, he was interested in establishing evolution as the solution to this complex design problem.[68] This conviction that Darwin had provided the true solution to the design problem, in the form of the theory of natural selection, animated the entire book. Dawkins's goal was to show that the slow, steady action of selection made the natural origins of complex life believable. Although it would be statistically improbable, even miraculous, for something like the human eye to appear fully formed in the natural world, Dawkins argued to his readers that they were not required to believe this about evolution. However, they could believe in a small series of small steps from one version of the human eye to the next.[69] For Dawkins, the smooth transition from one biological form to the next made evolution believable. By having natural selection work in tiny increments, there was no need to invoke the action of God.

Dawkins's conviction about the smooth action of natural selection put him in direct conflict with the public rhetoric surrounding Gould's "leaping" theory of punctuated equilibria. He devoted an entire chapter of the book, "Puncturing Punctuationism," to the subject.[70] Dawkins argued that the theory of punctuated equilibria was no different from traditional Darwinism and that perceived differences were the machinations of dishonest journalists—and perhaps some overstepping by Gould and Eldredge. He proclaimed "loud and clear . . . the truth: that the theory of punctuated equilibria lies firmly within the neo-Darwinian synthesis."[71] Dawkins argued that those who embraced the theory were actually as gradualistic as any other Darwinians; they merely "insert long periods of stasis between spurts of gradual evolution."[72] According to Dawkins, Gould and Eldredge did not really intend their theory to suggest a saltationist view of evolution; he believed that they had been seduced by the poetic rhetoric of suggesting that their theory was more revolutionary than Dawkins believed it actually was. Saltationism, in Dawkins's view, was the idea that evolution could proceed by jumps; there were hints in punctuated equilibria that it suggested the same. Dawkins could not put the point strongly enough: "This is no petty matter. In Darwin's view, the whole *point* of the theory of evolution by natural selection was that it provided a *non*-miraculous account of the existence of complex adaptation. For Darwin, any evolution that had to be helped over the jumps by God was not evolution at all."[73]

Dawkins contended that Gould and Eldredge were aware of these points, but he speculated that they had been persuaded by media sensationalism to associate their theory with saltationism. Dawkins further insisted that their theory might have been "modestly presented as a helpful rescuing of Darwin. . . . Indeed that is, at least in part how it *was* presented—initially."[74] As a "minor gloss on Darwinism," Dawkins asserted that punctuated equilibria did not merit such a "large measure of publicity."[75] The only explanation for such interest was that there were "people in the world who desperately want not to have to believe in Darwinism" and thus "if a reputable scholar breathes so much as hint of criticism of some detail of current Darwinian theory, the fact is eagerly seized on and blown up out of all proportion."[76] Gould and Eldredge, in other words, held some responsibility for the journalistic distortions of their theories merely because they dared to question the reigning interpretations of Darwin's theories.

Gould, predictably, rejected Dawkins's charge. He later wrote to the philosopher Michael Ruse, "I am angry with Dawkins because I have always taken him seriously (in strong opposition), while he tried to trivialize me by rhetoric in *The Blind Watchmaker*."[77] His opportunity to respond to Dawkins came two years later, in 1987, during a staged and, as Gould put it, much "ballyhooed" public debate hosted by the Darwin scholar and historian of science John Durant at Oxford

University. The debate was steeped in the trappings of evolutionary history; Dawkins and Gould were introduced as "the Huxleys of our age" and as the "two foremost representatives of evolutionary biology today." Although neither scientist offered anything particularly new in the debate, Dawkins's comments emphasized his view that adaptative explanations was not merely a useful model or a research heuristic. To him, it was the cornerstone of scientific naturalism.[78] Held at the Sheldonian Theatre at Oxford, just a few streets away from the sight of the famous 1860 debate between Thomas Henry Huxley and Bishop Samuel Wilberforce, the debate seemed to bring the conflict between Darwinism and Christianity into a new era.[79]

The debate between Dawkins and Gould was held shortly after the US Supreme Court handed down a decision on creation science in American public schools. Following precedents set in *McLean*, *Edwards v. Aguillard* (1987) determined that creation science was religiously motivated and therefore unconstitutional. For a brief moment, it appeared that the trial might have ended the political efforts of the creationist movement and perhaps the fight over Darwin's legacy within evolutionary science.[80] However, the dissipation of the creationist enemy was short-lived. In 1989 the director of the NCSE, Eugenie Scott, wrote to Gould with a terse assessment of a new situation: "We have a problem. A slick new creationist book, *Of Pandas and People*, was published in August, and has been making the rounds." Scott explained to Gould, "This is not ICR stuff; it's much more subtle. . . . The book makes a big push for the idea that 'intelligent design' does not require resort to the supernatural (honest)."[81] Several years later, Gould would describe *Of Pandas and People* as the "most 'sophisticated' (I really should say 'glossy') of creationist texts."[82] The new textbook manifested what Gould and Scott regarded as an alarming ability to reformulate a creationist perspective without the more conservative and extreme aspects of flood geology and creation science in the guise of an academic book.

The real salvo of this new, politically more astute movement came two years later, in 1991, with the publication of *Darwin on Trial* by Phillip Johnson. Johnson, a Harvard graduate and a law professor at the University of California, Berkeley, represented a much greater threat in the minds of many academics.[83] *Darwin on Trial* launched a full-frontal assault on the academic integrity of evolutionary biology. In it, Johnson put contemporary evolutionary theory "on trial," weighing the evidence for and against the theory in the manner of a legal briefing. His conclusion—that Darwinian evolutionary theory was wanting—was based on legal, not religious, credentials. Johnson wished to create a space for discussion in the face of what he believed was the hegemonic hold science had over public culture: "If we say that naturalistic evolution is *science,* and supernatural creation is *religion,* the effect is not very different from saying that the former is true and the latter is fantasy."[84] Thus, he devoted his book to detailing the disagreements, controversies, and faults he perceived concerning the edifice of evolutionary science.

Johnson's book had a bigger target than the philosophical and interpretive differences within evolutionary biology. He also charged scientists with excluding others from their debates. They had set the terms of their own conversations, and they would not allow outsiders into their circles. His criticism of Gould in a chapter titled "The Fossil Problem" particularly reflected this: "Paleontologists seem to have thought it their duty to protect the rest of us from the erroneous conclusions we might have drawn if we had known the actual state of the evidence. Gould described 'the extreme rarity of transitional forms in the fossil record' as the 'trade secret of paleontology.'"[85] Johnson attempted to portray Darwinism as a frayed consensus, a theory held together only by the stubborn dogmatism of evolutionary biologists, who refused to acknowledge their own disagreements.

This strategy of framing evolution as a "theory in crisis" was also the mandate for the Center for the Renewal of Science and Culture, a project of the nonprofit Discovery Institute. Founded in 1996 as the institutional heart of intelligent design, the center put together a series of "teach the controversy" campaigns, which argued for the teaching of intelligent design as a scientific alternative to Darwinian evolution in American public schools.[86] The Discovery Institute's desire to use intelligent design as a "wedge" through which to reform American culture according to Protestant Christian values was made public when the infamous "wedge document" outlining this strategy was published on the institute's website in 1997.[87] The Discovery Institute cast academic Darwinism as its foe in a war over values in American culture.

The emergence of the intelligent design movement meant that creationism remained an ever-present backdrop for those who busied themselves with keeping Darwin's legacy in the final decades of the twentieth century. Earlier generations of American biologists had looked to Darwin to unify evolutionary biology. For Dawkins and Dennett's generation, Darwin's greatest triumph would come in his final defeat of religion. In 1986 Dawkins had been willing to give Gould the benefit of the doubt, laying the blame for associating Gould's criticisms and creationism primarily at the feet of errant journalists. But faced with the new threat of intelligent design, these Darwinists closed ranks. This was particularly clear in their ongoing engagements with Gould's criticisms of adaptive explanations.

In *Darwin's Dangerous Idea* (1995), Dennett argued that Gould's criticisms of adaptive explanations in evolutionary biology were overblown, misguided, and false. Dennett described himself as having been initially astonished to find so many of his colleagues in the humanities convinced by Gould that Darwinian theory had been overthrown. After some reflection, Dennett realized that Gould's campaign against the modern synthesis over the two decades since the publication of "The Spandrels of San Marcos" had contributed to the academic and "public misconception that evolution was dead." But Dennett was delighted to report that "no sooner has he [Gould] gone into battle than the modern synthesis has shown its flexibility, readily absorbing his punches, to his frustration."[88]

The title of Dennett's book foreshadowed the major thrust of his argument: that Darwin's theory of evolution by natural selection was a revolutionary idea, in fact *the* revolutionary idea; evolution was simply an algorithmic process. Dennett described Darwin's contribution to human thought in grand and harsh terms: "If I were to give an award for the single best idea anyone ever has had, I'd give it to Darwin, ahead of Newton and Einstein and everyone else. In a single stroke, the idea of evolution by natural selection unifies the realm of life, meaning and purpose with the realm of space and time, cause and effect, mechanism and physical law."[89] This was a Darwin of grand revolution—of dangerous purpose—who had brought humanity out of the clouds of delusion and forced us to face the cold, hard facts of our purposeless existence. Darwin here was a towering genius, above even Newton and Einstein. This was not the Darwin that Gould recognized from his participation in Darwin studies—puttering about with earthworms, exchanging letters with Huxley, or having doubts about his status aboard the *Beagle*. Darwin had been put on trial by the creationists, and these Darwinists responded with a fearsome man of towering intellect.

What did Dennett mean when he described evolution as an algorithmic process? Dennett argued that the anachronistic term captured the essence of Darwin's theory. As Dennett explained, "An algorithm is a certain sort of formal process that can be counted on—logically—to yield a certain sort of result whenever it is 'run' or instantiated."[90] Dennett argued that design was not sculpted into the world by the hand of God; rather, it accumulated through the continual action of natural selection. The complexity and ability of God were, in a real sense, the product of natural selection; selection churning over millennia had slowly built life, then complex life, then intelligent life, and then humans with the cognitive capacity to imagine an intelligent designer. Although humans were all continually tempted to give explanations for biological phenomena that required a lifting from above, Dennett believed that we would only truly understand the world and ourselves by recognizing the algorithmic nature of natural selection from below.

From Dennett's perspective, the conflation of Gould's criticisms of adaptive explanations in Darwinism with creationism was not ultimately the fault of journalists nor even of creationists themselves. Rather, Dennett contended that the fault lay with Gould himself because of what Dennett framed as a frightening truth: "America's evolutionist laureate has always been uncomfortable with the fundamental core of Darwinism."[91] Dennett concluded that Gould was harming Darwinism because he himself was not truly a Darwinian. Dawkins had argued that the canonical tenet of Darwinism was its answer to adaptive complexity; Dennett now confirmed that the "universal acid" of this truth must be accepted by those who claimed to be Darwin's heirs.

Scientific Failure: The Case of the Female Orgasm

By the end of the 1980s the new creationist threat had overshadowed the earlier public debates regarding sex difference and scientific racism. The clearest sign of this transformation was new evolutionary images that materialized in American consumer and media culture after *McLean*. Particularly in depictions of Darwin, themes of science and religion eclipsed debates over race and gender. A robust panoply of Darwiniana proliferated in this period: Darwin's portrait, images from his notebooks, and quotations from *On the Origin of Species* adorned t-shirts, mugs, and websites.[92] Practicing biologists tattooed these figures onto their bodies in shows of solidarity with the Darwinian worldview.[93] The Darwin fish faced off with the Christian ichthys on car bumpers across the nation.[94] However, the popularity of this imagery was not a sign that evolutionary approaches to sex difference had waned during this period. The adaptive logic of sociobiology continued to mold American imaginaries of masculinity, sexuality, and Western identity. But since leftist critiques of sociobiology were largely academic, it was the debate between creationism and evolution, not feminists and sociobiologists, that dominated the American public imagination.

However, the sociobiological account of sex difference continued to hold sway over biological research in the 1980s despite the ongoing critiques of feminist science studies. But the logic of adaptation that had constructed sex difference in *The Selfish Gene* and *Sociobiology* now denied supernatural explanations for life. When Gould began working with the feminist philosopher Lisa Lloyd in 1986 to critique sociobiology's approach to sex difference, sociobiologists interpreted it as a call to give up on science itself.

Gould met Lloyd when she came to Harvard to study in the Department of Genetics with Lewontin, and he would later publish with Lloyd on the debate over levels of selection.[95] A philosopher by training, Lloyd had been gathering evidence on the insufficiency of adaptive explanations for the female orgasm. She later published this research in her 2009 book *The Case of the Female Orgasm: Bias in the Science of Evolution*.[96] In the book, Lloyd argued that the sociobiological commitment to an adaptive explanation for the female orgasm was unwarranted by the evidence. After examining twenty accounts for the female orgasm as an adaptation, Lloyd contended that it was best understood as a by-product, as a phenomenon homologous to the male orgasm (which had a clear adaptive function). Lloyd insisted that setting aside an adaptive explanation was not synonymous with abandoning an evolutionary account for the female orgasm. Gould liked Lloyd's work. He used it as the basis for a column early in 1987 titled "Freudian Slip" (renamed "Male Nipples and Clitoral Ripples" in the Norton anthology), in which he discussed the evolution of male nipples as well as that of the female orgasm.[97] Gould argued that both should be understood as products of the homologous nature of sex differentiation rather than as adaptations in their own right.

For the column, Gould pulled material from Lloyd's ongoing research, seeing it as an ideal example of the misguided emphasis on adaptation in evolutionary research. He began the piece by ruminating on a popular confusion about the evolutionary process that had appeared in "hundreds of letters" he received "from readers." Focusing on one letter, he related the writer's bewilderment to the rest of his readers: "Why do men have nipples? . . . This question nags at me whenever I see a man's bare chest!"[98] Noting the inadequacies of the proposed answers to the reader's question—that men nursed or that humans had once comprised only one sex—Gould stated that the mystery would not be resolved by additional research. No, he argued, "as with so many persistent puzzles, the resolution does not lie in more research within an established framework but rather in identifying the framework itself as a flawed view of life." This flawed framework was the "commitment to pervasive utility for all parts of a creature."[99]

From Gould's perspective, insisting on an adaptive function—a benefit given to males by their nipples, or to females by their orgasms—bordered on the absurd. He contended that "males and females are not separate entities, shaped independently by natural selection." Since both sexes were "variants upon a single ground plan, elaborated in later embryology," the fact that females needed nipples was enough to explain why males had them. A developmental explanation was as legitimate an evolutionary explanation as an adaptive explanation.[100]

Not only was it a legitimate scientific explanation but to Gould it was a fuller, more complete view of the evolutionary process.

Gould's attack on adaptive explanations elicited an immediate response from the sociobiological community. John Alcock, a leading behavioral ecologist whose empirical research focused on the adaptive value of male mate choice, wrote a letter to the magazine's editor that was published in the subsequent issue of *Natural History*. Alcock took Gould to task, accusing him of writing "just how" stories "as part of a plea for a pluralistic evolutionary biology, one that promotes a more complex, and more realistic worldview than that of the ardent adaptationist, who insists on searching for the reproductive value of every evolved character." Alcock acknowledged that "the pleas for pluralism seem reasonable and wholesome." Why, then, Alcock asked, were so many adaptationists so hostile to them? Because, he suggested, "the adaptationist position is an invitation to scientific investigation."[101]

Alcock's response highlighted a crucial element in the sociobiological perspective, that is, that Gould's arguments took questions about the female orgasm outside the realm of "scientific investigation." Alcock's letter revealed the belief that adaptive explanations were not one among many possible types of evolutionary accounts. Rather, the "adaptationist position" was "an invitation" to science itself. The implication was that accounts that criticized the adaptationist position rejected science itself.

A decade later, Alcock decided to "tilt at the windmill of Stephen Jay Gould" by lambasting Gould in the pages of the journal *Evolution and Human Behavior* in a piece titled, "Unpunctuated Equilibrium in the *Natural History* Essays of Stephen Jay Gould."[102] Alcock expressed his dismay that Gould's *Natural History* platform allowed him to carry on "unfazed, repeating to a large general audience the same arguments that his colleagues in evolutionary biology have examined, refuted, and dismissed long ago." Alcock again asserted that Gould's antipathy to adaptationism was not really a plea for pluralism but rather a deep prejudice that inhibited the practice of evolutionary biology. Surveying the fields of sociobiology, evolutionary psychology, and the pages of *Natural History*, Alcock declared that adaptationism was "not only alive but flourishing."[103] In his view, this was entirely reasonable, since "adaptationists" made "use of an extraordinarily powerful tool: Darwinian natural selection theory."[104] Alcock would go on to express his view of the crowning achievements of adaptationist biology in his aptly titled book *The Triumph of Sociobiology* in 1999 [105]

But those criticisms were yet to come. In 1987 Gould had the opportunity to respond to Alcock, and his rejoinder was printed alongside Alcock's original letter in the same issue of *Natural History*.[106] Gould's zeroed in on Alcock's letter as a demonstration of the restrictive nature of the current evolutionary paradigm. He asserted that Alcock had "demonstrated beautifully, if unconsciously, the force and restrictiveness of adaptationist bias."[107] Arguing that Alcock and other sociobiologists incorrectly equated "current utility with reasons for historical origin," Gould reiterated his argument that the empirical evidences of history—either in the fossil record or, in this case, in developmental pathways—should alter biologists' assumptions about the nature of the evolutionary process. Rather than understanding such human cultural achievements as reading and writing as the products of the algorithmic process of selection searching in an empty design space, Gould saw human intelligence and complexity as the mere happenstance of history. They were the "nonadaptive structural consequences of a large brain, built by natural selection for other reasons."[108]

For Alcock, explaining human behavior as the result of a selective process that sought reproductive utility was the hallmark of good science. Gould believed just the opposite. In Gould's view, it was the openness to more than one kind of evolutionary explanation, not the exclusive focus on one type, that characterized "good science." He stressed that this "good science" needed "testable hypotheses of all feasible types—including developmental and adaptationist."[109] Gould undoubtedly saw his writing as opening up the space for alternative evolutionary explanations. Given the vehemence and persistence of the responses from sociobiologists, however, it is arguable that Gould's persistent and vocal excoriation of adaptive explanations helped to clarify and solidify the opposing position.

At first glance, these debates over the evolution of sex organs appear to be far removed from the public contests between evolutionary researchers and the American creationist movement. However, the creationism controversy significantly heightened the explanatory importance of adaptation. In the face of threats from creation science and intelligent design, evolutionary researchers held even more tightly to the power of adaptation. It is not my contention that researchers such as Alcock put forward adaptive explanations for contemporary gendered behavior in a deliberate effort to combat the epistemological and political threat of creationism. But in Alcock's argument that other evolutionary biologists had "consistently debunked" Gould, he pointed to texts such as *The Blind Watchmaker* and *Darwin's Dangerous Idea*, which explicitly offered Darwinian adaptation as proof against the argument from design and thus a defeat for religious belief.[110] During the 1980s and 1990s, adaptation as a research heuristic and adaptation as the cornerstone for scientific naturalism became inextricably intertwined.

During these years, Darwin's image moved in spaces and registered with audiences beyond the reach of these professional debates. For every creationist author who carefully watched the technical disagreements among scientists, there were far larger audiences whose acquaintance with the creationism debate was through televised debates, megachurch leaflets, or local campaigns waged to introduce creationism into public classrooms. In the years when the Christian Right was first ascendant, creationism became an intuitive component of the new evangelical political identity.[111] This meant that the slippages between creation science and Gould's heresies against neo-Darwinism did connect with existing audiences. But the details of his theories, or his provocations against the synthesis, did not animate larger political debates. When Republican politicians questioned Darwinian evolution during election seasons, they did not get into the finer details of natural selection. Publicly doubting evolution was a way to align with voters on an entire political and religious worldview, one that was committed to the heterosexual nuclear family and took pride in American exceptionalism.[112] This had several consequences. Supporting the veracity of evolutionary science became a token of liberal politics by the turn of the twentieth century. At the same time, leftist perspectives on science largely moved into the academy and remained on the edges of public debates. This was particularly true in the case of feminist discussions of sociobiology and evolutionary psychology.[113] And as this happened, Gould's version of Darwin faded into the background of the public imagination. The altercations between Darwin and creationism were far more compelling intertwined in the nation's cultural and political debates. Although evolutionary accounts of sex, race, and gender did not abate, these years cast Darwin's revolution into a war of cosmologies.

Criticizing E. O. Wilson

When Alcock complained that Gould's explanatory pluralism left behind scientific investigation or Wright accused Gould of being an "accidental creationist," both revealed a set of entwined intuitions. First, that the theory of natural selection—and with it, the person of Darwin—was the foundation of a Western secular identity. And second, that any specific criticisms of this theory by the academic Left did real harm to the authority of Western science. These positions bring us to a central question animating this book: Did the leftist criticisms of science that first emerged in the radical science movement damage the public authority of science in American society? Or perhaps more importantly, should we regard criticizing science for its historical racism and sexism as primarily a move to halt knowledge creation about the natural world? One way to reflect on this question is through a set of historical criticisms of Wilson and *Sociobiology* that came to light several decades after these exchanges.

In December 2021 Wilson passed away at the age of 92, after a decades-long career as an evolutionary scientist, naturalist, and prolific author. His death was memorialized by many in the scientific community, who praised his research across many fields of evolutionary biology and his public advocacy for environmental conservation.[114] But one piece took a more critical approach to Wilson's scientific legacy. In an essay published in *Scientific American* a few days after Wilson's death, the reproductive health scientist Monica McLemore argued that we "must reckon with [Wilson's] and other scientists' racist ideas if we want an equitable future." McLemore's primary criticism of Wilson was the role she claimed *Sociobiology* had played in contributing to the false dichotomy of nature versus nurture in the study of human social issues. Using this point as an opening to a broader critique of racism in contemporary medicine and science, McLemore contended that a range of strategies can and should be employed to engage with the problematic work of scientists whose "legacy is complicated." These include "truth and reconciliation" about citational practices, "diversifying the scientific workforce," and developing "new methods" that are sensitive to the legal and ethical implications of empirical science.[115] According to McLemore, only through the hard work of confronting the history of scientific racism can the scientific community—and the wider public—create a just world for all.

Following this article, two of my colleagues in the history of science, Mark Borello and David Sepkoski, published a piece titled "Ideology as Biology" in the *New York Review of Books*. Writing for the same venue that had published the SSG's original criticism of sociobiology almost forty years earlier, Sepkoski and Borello added to the discussion of Wilson's legacy in the manner of "good historians" by delving into Wilson's archive to weigh in on the influence of racial bias on his science. What they recovered was "a lengthy and revealing correspondence" with J. Philippe Rushton, a "notorious race scientist," in which Wilson "more openly associated his own scientific ideas with racialized views of human ability than he ever did publicly." In their piece, Borello and Sepkoski detailed Wilson's substantive intellectual and institutional support for Rushton's suggestion that one of Wilson's ecological equilibrium models, r/K selection theory, could be used to explain the social status of Black Americans.[116]

Wilson had developed the r/K selection theory in the 1960s with the population biologist Robert MacArthur as a way of modeling reproductive strategies of different species of animals.[117] The model proposed that animals adopted one of two strategies emphasizing either quantity or quality in their offspring. An example of an *r-selection* group is mice, which have many offspring, exhibit less parental care, and have a relatively short time to sexual maturity. Animals like whales and elephants follow a *K-selection* strategy, with longer life expectancies and fewer offspring, who require greater care from the parents while they are young. Although the r/K selection model eventually fell out of favor, it remained a dominant heuristic for many decades after Wilson and MacArthur published their work.

Rushton suggested that this model ought to be applied to the difference between Black and white Americans; he argued that the former were "r-selected" and the latter "K-selected." Not only did Wilson agree with Rushton's views in their personal correspondence, calling it "one of the most original and interesting [ideas] I've ever encountered in psychology" and describing it as "courageous," but he also supported Rushton's (eventually unsuccessful) efforts to have these views published in the journal *Ethology and Sociobiology*.[118] As Borello and Sepkoski recount in their history of these exchanges, drawing out the blatant racism of Rushton's arguments and Wilson's support for them requires little interpretive work. We only have to recognize the danger of comparing any so-called race to animals generally considered vermin (i.e., rodents), as well as the fact that the model was meant to compare separate biological *species*, not genetically indistinguishable cultural groups within the same species. That is, Wilson attempted to aid Ruhston's efforts to use evolutionary theory to prove that white and Black Americans were different enough to be modeled as separate species and that these biological differences ultimately drove the racial social hierarchies of American society.

I admit that I found the history detailed by Borello and Sepkoski truly surprising. Over the course of my research, I mostly encountered sociobiology as a field more invested in understanding sexual differences than in understanding racial ones. For the most part, race is implied, rather than discussed, in sociobiologists' technical models of evolution. Wilson himself was careful not to directly address racial differences in his public writings after the SSG accused him of racism during the 1970s.[119] The gravity of what Borello and Sepkoski have brought forward about Wilson's involvement with Rushton's work reminds us of the lasting power of forms of leftist criticism based in moralized public shaming.

It is my belief that scientific racism is better understood as structural violence than as a question of individual morality. For these reasons, in earlier chapters I focused on some of the limitations of critiques of personal bias in science enacted by Gould and other radical scientists. Gould's view in *The Mismeasure of Man* was that this type of confrontation with racial bias could prompt the individual and collective self-reflection necessary to create a better science. As we have seen, this was both a specific kind of political argument about how social change occurs and a particular understanding of the nature of scientific knowledge in relation to that social change. In this book, I am interested in working through the political possibilities of other ways of conceptualizing the relationship between science and democratic futures. And yet: it remains necessary to have tools to rebuff the cultural authority of science when it is used to support racial stereotypes and further racist social policies. The hope that Gould expressed in *The Mismeasure of Man*—that exposing the history of racism in science can help move us forward toward a more equitable future—is threaded through the pieces by McLemore and Borello and Sepkoski.

Another aspect of this episode illustrates how the criticism of science can be a generative and not solely a corrective exercise. As it turns out, Borello and Sepkoski partly organized their history in direct response to the pushback from within the scientific establishment to McLemore's *Scientific American* article. A group of scientists, including several members of the National Academy of Sciences, signed an open letter in rebuttal to McLemore's piece on a Substack newsletter authored by the science blogger Razib Khan.[120] In their discussion of the episode, Borello and Sepkoski highlighted Khan's own history of writing for alt-right and white nationalist publications and observed that the alliance between the scientific establishment and "fringe proponents of race science" was all the more reason to delve into Wilson's legacy. But although Khan embraced Western identitarianism more explicitly than other scientific liberals, he expressed a view of leftist and especially feminist critique—that it functions primarily to silence and inhibit the scientific quest for truth—that was shared by many other liberals, including Wilson, from the sociobiology debate onward.[121] David Sloan Wilson, an evolutionary biologist and public intellectual, posted a similar response denigrating McLemore's scientific understanding. He described McLemore's piece as "bewilderingly flimsy" and as "demonstrating a baffling ignorance."[122] What emerges from these responses is the liberal position honed after many decades of culture war with the Left, a position that resisted the Left's representation of science's historical sexism and racism and presented science as an embattled quest for truth.

But these criticisms have always been hollow. As we have seen, Gould's objections to sociobiology were always as much about revolutionizing the field of evolution as about criticizing the specific models of sociobiology. By exploring in detail Gould's technical criticisms of sociobiology alongside his political understanding of scientific racism, I have hoped to highlight how the leftist critique of science could be generative as well as corrective. The same can be said for the scholars who developed feminist, critical race, indigenous, postcolonial perspectives on science. These were not paradigms built to silence the robust engagement of ideas or earnest attempts to understand the natural world. They were declarations that different people with different bodies saw and felt and understood the world differently than did the idealized knower of the Western Enlightenment.

I believe that rather than seeing these critiques of science as threats to knowledge creation, we should see them as gifts to one another. After all, Wilson himself had a unique way of viewing the world that was the direct result of his own embodiment. His 1994 memoir *The Naturalist* tells the story of a fishing accident in his youth that damaged his right eye, causing him to lose his stereoscopic vision. As a teenager he also lost his hearing at higher registers, making it difficult for him to hear birds or frogs. Reflecting on these incidents in later years, he argued that he "was destined to become an entomologist . . . not by any touch of idiosyncratic genius . . . but by a fortuitous constriction of physiological ability."[123] The particularity of Wilson's way of seeing the world helped him to birth a new evolutionary discipline. Was this not also the case for the feminists, critical race theorists, and indigenous writers who built their theories in the decades after 1975?

Is a Democratic Science the End of Western Secularism?

> To which, let all people of goodwill; all who hold science, or religion, or both, dear; all who recognize NOMA as the logically sound, humanely sensible, and properly civil way to live in a world of honorable diversity, let them say, Amen.
>
> —*Rocks of Ages: Science and Religion in the Fullness of Life, 1999*

Gould passed away in the spring of 2002, less than a year after the attacks on the Twin Towers in New York, an event that irrevocably transformed American public dialogue on science and religion. After September 11, 2001, many Americans imagined themselves in a historic battle between Western civilization and the dogmatic religious superstition of a militarized Islamic world.[1] This narrative was propagated and circulated by the George W. Bush administration, the American news and entertainment media, and the writings of the evangelical establishment. It was also galvanized by the world of Darwinian celebrity. Two of Gould's sharpest public critics, Dawkins and Dennett, became one-half of the foursome who became known as the "four horsemen of New Atheism," each of whom published a book in the decade after the fall of 2001 denouncing the influence of religion in public life.[2] Along with Samuel Harris and Christopher Hitchens, Dawkins and Dennett argued that religion was not simply misguided but a threat to an enlightened society. Although each of these authors had longstanding disagreements with religious belief, the events of September 11 prompted them to cast these criticisms in the language of civilization's rise and fall.

After September 11, the New Atheists, as the four horsemen and other like-minded public intellectuals came to be called, transformed Darwin into the patron saint of Western secularity. In their view, *The Origin of Species* had ushered in the final stages of Western secularism, which had begun with the Enlightenment. They would declare the scientific humanism of the Enlightenment "the de facto morality of all modern democracies" and the foundation upon which equality and liberty were based.[3] Wilson's intuition during the sociobiology controversy—that science ought to be defended and not critiqued—was given a clearer and bolder life in a world where religious dogma represented more than the teaching of erroneous geology in southern classrooms. In their public writings and speaking, these intellectuals argued that religious faith mired cultures and communities in the past. They found parallels in what they viewed as the ignorance of American creationists and Islamic terrorists. And they argued that the only way forward was a relentless intellectual bravery that set aside the emotional crutch of supernatural belief. According to Dawkins and Dennett, only "brights" were able to face the cold reality of the material universe without flinching.[4]

Scholars writing from the perspective of the critical humanities questioned the cultural assumptions at the heart of the New Atheism. In 2003, the postcolonial studies scholar Talal Asad asked whether secularism should be regarded as something that is "necessary to universal humanism, a rational principle that calls for the suppression . . . of religious passion so that a dangerous source of intolerance . . . can be controlled, and political unity, peace, and progress secured," or, rather, as a "colonial imposition" that engendered violence. Asad concluded that Western secularism was not a transcendent worldview but instead a form of governance that dismissed the humanity of certain people.[5] The evidence supporting his conclusion can be seen in the long wake of the United States' military intervention in the Middle East, its disregard for human rights at Guantanamo, and its policies of detention and policing of Muslims across and within its borders. The view that secularism is necessary for democracy reveals itself in practice to be an argument for the cultural supremacy of the West as a racialized and religious order.[6] After all, even the most strident of the New Atheists were more comfortable with the cultural forms of their own religions than with Islam.

What, then, is the relationship between secularism and a democratic society? When leftist activists and scholars began working to create radically inclusive practices for the creation and use of science, whether in universities or in leftist health collectives, their intention was not to diminish the cultural authority of science to make room for religion. They built a critique of science that remains one of the most powerful avenues for reckoning with modern science's history of patriarchal, colonial, and racist violence. Nevertheless, critical race and feminist critiques of science did become entangled in the creationist attack on the public authority of Darwinian science in the final decades of the twentieth century. We must thus ask whether leftist critique functions only to remove the biases of science or supports wholly different ways of knowing, including religion, in the public sphere. And if it does support religion, is this a problem?

Gould's answer to this question was developed in his most famous—and famously criticized—philosophical account of the relationship between science and religion. In his 1999 book *The Rocks of Ages: Science and Religion in the Fullness of Life,* Gould argued that the proper relationship between science and religion was best described by the phrase *nonoverlapping magisteria,* or NOMA. Gould borrowed the term *magisteria* from Catholicism, where it refers to the church's authority to establish the true and correct teachings of the faith. Gould proposed that science and religion occupied separate (but equally important) magisteria, whose relationship should be regulated by a "principle of respectful noninterference."[7] According to him, science covered "the empirical realm," whereas religion "extends over questions of ultimate meaning and moral value."[8] Gould contended that by recognizing this division, well-informed persons could draw on both religious tradition and scientific knowledge in their pursuit of a fulfilling and moral life.

Gould's NOMA was debated by scientists, theologians, and scholars across the academic world almost from the moment of its publication. In the subsequent decade, most scholars who assessed Gould's proposition for the relationship between science and religion found NOMA impractical and shortsighted.[9] What has not entered the public debate over NOMA is that Gould's framework for science and religion emerged out of the same paradigm that had shaped his critique of scientific racism in *The Mismeasure of Man.* Gould's argument for the importance of valuing religion and the humanities reflected his desire to have an epistemological counterbalance to the authority of science that could contend with its historical racism. In his private papers, Gould made explicit the connection between his pluralistic approach to the authority of science and religion and his vision for American democracy, one in which a multitude of faiths, races, and creeds would come together in a society built on tolerance and mutual goodwill.[10]

Situating Gould's proposal for NOMA within its political milieu helps us to see how it functions as an alternative to the vision offered by scientific liberals and the New Atheists. It embraced the role of other ways of knowing to contend with the violence of science, while simultaneously imagining a multifaith and multiracial American democracy. Gould built this alternative during the years when evolutionists' public advocacy for their version of Darwinian evolution increasingly rejected feminist and critical race accounts of science in efforts to fortify the authority of science against American creationism. This chapter traces the expansion of a reactionary scientific liberalism during the years of the academic science wars and the rise of the New Atheists. Revisiting Gould's NOMA through the lens of the leftist attempt to build a democratic science helps us to connect debates over secularism and American democracy with debates over the critique of scientific racism and sexism.

But let us consider the significant limitations of NOMA as a philosophy. Gould built the theory on a Eurocentric definition of religion. NOMA, as such, does not contend with the material entanglements of science and religion that shape the most pressing debates over reproductive technologies, health care, climate change, and human migration crises. My call here is not for the adoption of Gould's NOMA as the final answer to the relationship between science and religion. Rather, I suggest that entering a conversation with NOMA during the period in which it was formed allows us to see why liberals viewed leftist work to build a democratic science as a threat to secularism and, further, how secularism became enshrined by liberals as the necessary precursor to democracy and social equality. In my view, we do not need to fear the end of secularism in the work for a radically democratic science. With intellectual humility, we can embrace profoundly different, even contradictory, ways of knowing and being in confronting the world's most challenging problems.

The Science Wars

Long before the emergence of the New Atheism, cultural commentators had worried over the state of intellectual life in the West. In his famous 1959 Rede lecture, "The Two Cultures," C. P. Snow decried the gulf then developing in the academy between the humanities and the sciences. Snow's lament arose from his specific experiences at Cambridge University, an institution that was experiencing rapid change in post–World War II Britain.[11] A similar split was developing in the United States as Big Science transformed the federal government, research universities, the nation's capacity for global military action, and its imperial ambitions.[12] What was the role of the humanist in such a capaciously technical world? Even Snow could not have anticipated how deep the division between the sciences and the critical humanities would become by the final decades of the twentieth century. When he delivered "The Two Cultures," it was possible for him to argue that scientists and humanists in the West were "comparable in intelligence, identical in race, not grossly different in social origin," and that their disciplinary training was responsible for the differences in their "intellectual, moral, and psychological climate[s]."[13] The new interdisciplinary fields that emerged on university campuses in the following decade contradicted this assertion.

In the 1970s, new bodies bearing new identities moved into the academy. New departments and programs in such fields as women's, critical race, postcolonial, queer, and multicultural studies led to a new emphasis on the relationship between identity and scholarship. The campus politics of the New Left era were formalized and concretized into new practices and epistemologies of academic inquiry. Practitioners of these disciplines questioned whether knowledge should change if new bodies made it, and they reframed the proper social role of the academy as one of social justice. If, as these scholars argued, elite knowledge had been used for centuries to categorize and oppress the disenfranchised, it was not enough for these bodies to be merely represented in the university. Rather, these "critical humanists" called upon academics to understand knowledge production as inescapably and relentlessly political.

The ethical claims of these scholars had different goals and were intended for different audiences, than evolutionary biologists' commentary on the relation between knowledge and politics. Biologists saw creationism as both a social and an epistemological threat, but one that largely operated outside the university gates. Critical humanists, on the other hand, viewed their ethical task as being at least partially centered within the academy itself. Their critiques were directed at the practices, institutions, and epistemologies of the modern university, an extension of the leftist belief that knowledge itself was politically transformative. By the end of the twentieth century, the differences between the "intellectual, moral, and psychological climate[s]" of the critical humanities and the sciences were no longer merely an issue of disciplinary training.

Of course, not every academic who was gender-nonconforming, queer, or of non-European descent was to be found in these new interdisciplinary spaces. Nevertheless, a demographic and epistemological change began to occur in the infrastructure of American universities in the late 1970s and continues to this day. At the heart of the question of the relationship between science and democracy was another profound question: who can do science? Liberal scientific intellectuals asserted that science was universal knowledge that stood apart from, but could inform, political ideologies. In their view, anyone who was willing to follow the methodological standards and professional conventions of the scientific community could do science. By contrast, critical humanists argued that the authority of scientific knowledge was embedded not in its methods but in the historical authority of elite Western masculinity. They contended that there was no apolitical core to science; furthermore, they argued that science was responsible for sexism and racism, not because of the infections of social bias but because the structure of science itself produced those evils.

One way to understand this position is to consider the role of identity in the rise of modern science. In the early modern period, the validity of natural knowledge was dependent upon witnessing; the person who performed a natural experiment or observation could not bring the entire polity with them to the experience, and so they personally testified to the veracity of their claims about the natural world.[14] In this context, it was essential that the individual naturalist or natural philosopher be reliable since the veracity of their empirical claim depended on the trustworthiness of their character. Thus the truth of scientific knowledge was invested in particular bodies, specifically the bodies of the gentlemanly elite, whose testimony could be trusted.[15]

When the institutions and practices of the natural sciences expanded in subsequent centuries, these assumptions of gentlemanly character came under considerable scrutiny. In particular, it became necessary for the emerging institutions of natural science to formally conceptualize the trustworthiness of natural knowledge, even if the persons conducting observations and experiments were paid to do this work.[16] Professionalization was meant to create a space in which new bodies could become trustworthy actors in the production of scientific knowledge. The refining of scientific methods and the standardization of disciplinary credentialing helped expand the class origins for men of science in the Anglophone context during the nineteenth and twentieth centuries.[17] Scientific training promised that men from any class background could be scientists; if they disciplined themselves properly, they would produce scientific knowledge that was universal, timeless, and objective.

There is a grand hopefulness in this confidence in the promises of a universal science, but the process of professionalization was not truly liberatory. Certainly, it allowed more bodies into the circles of science, but even as men argued that more classes should have access to scientific knowledge-making, they used science to claim that Euro-American women and non-Europeans of any gender were physiologically incapable of intellectual work.[18] In a sense, science mirrored the broader European Enlightenment. Even as they claimed authority from a European class hierarchy and declared that all humans were "endowed with certain inalienable rights," Enlightenment figures on both sides of the Atlantic justified enslavement and denied suffrage to women and non-Europeans by arguing that those bodies were insufficiently human. Well into the twentieth century, these men of science turned the very tools of science—empiricism, quantification, standardization, observation, and categorization—into claims that other bodies could never fully participate in the life of the mind.

These injustices preoccupied a generation of white feminists and Black intellectuals in the United States at the beginning of the twentieth century. These scholars and activists recognized that scientific knowledge itself impeded their access to the world of science, and they responded to this discrimination in various ways. White feminists combated the masculine assertion of female mental deficiency, whether on the basis of menstrual periodicity or of small cranial size, by contending that unbiased medical knowledge would validate the intellectual equality of white women and men.[19] They blamed sexism for corrupting the practice of medical science and grounded their feminist activism in the promises of unbiased scientific validation. These women wrote treatises that criticized Darwin's theory of sexual selection, and they left their brains to medicine with the confidence that science would verify their intellectual equality.[20] Early-twentieth-century Black intellectuals considered the role that scientific knowledge played in preventing Black men and women from succeeding in the scientific profession. As we have seen, in 1939 W. E. B. Du Bois argued that social circumstance, not biological insufficiency, hindered Black Americans from excelling in science.[21]

The late 1970s, in other words, was not the first time that disenfranchised communities objected to the prejudices of science and American intellectual life. But the interdisciplinary scholars of the late twentieth century assembled a novel set of challenges to the relationship between scientific knowledge and liberal politics. The generations before them had argued that science held the promise of liberation if only it could be freed from the biases of social bigotry. These new intellectuals argued that science itself produced racism and sexism through the epistemic structure of those knowledge claims.[22] Postcolonial theorists in science studies were particularly articulate on this point, demonstrating the imperialism and extractive violence of the natural sciences during the modern era.[23] And perhaps, just perhaps, these new fields claimed, there could be other forms of privileged knowing besides science to which society would have to cede authority. The knowing of indigenous persons. The knowledge of women. The epistemology of Black Americans. With each field named for an identity, scholars argued that not only were these new bodies worthwhile but they possessed an authority that did not require the verification of science.

By the early 1990s, the leftist critique of science was sufficiently influential in the American academy that it provoked high-profile responses from a group of scientific liberals. These academics, largely from the natural sciences, reflected the intuition that had shaped Wilson's public and private stance during the sociobiology controversy, that is, that science was a wellspring of, rather than an impediment to, the flourishing of democracy. In a series of conflicts in both academic and popular publications, the science wars pitted liberal and leftist views of science against one another.

These intellectual contests are often dated to the 1994 publication of *Higher Superstition: The Academic Left and Its Quarrels with Science*, authored by the biologist Paul R. Gross and the mathematician Norman Levitt. Gross and Levitt published the book as an attempt to characterize the views and activities of what they defined as "the academic left" and "postmodernism." They explicitly positioned the text as a warning to the scientific community. In particular, they highlighted what they saw as postmodernists' disdain for the foundations of Western civilization, including the belief that the "intellectual schemata of the Enlightenment" was nothing "but a skeleton remains."[24] In the view of Gross and Levitt, postmodernist and leftist critiques of science had not made the academy more just and equitable, nor had they achieved a more democratic science. Rather, they argued that by critiquing science the academic Left had attempted to undermine one of the most important cornerstones in the edifice of Western democracy.

In his 1999 follow-up volume, *Prometheus Bedeviled: Science and the Contradictions of Contemporary Culture*, Levitt asserted that science was an essentially "selective" institution and that "no society, however egalitarian, has ever eliminated the sense that science is an elitist calling." Moreover, Levitt argued that even within science the most important innovation "comes from the work of a small number of extraordinarily able individuals." Levitt acknowledged that science's fundamental elitism might seem to be in conflict with democratic egalitarianism, but he nonetheless asserted that science had in fact "been historically associated with a democratic . . . mood."[25] He continued, "Both viewpoints arose and became entrenched, more or less, in the same place—Western Europe—at approximately the same time—somewhere between 1600 and 1800."[26] Why would this be the case? Levitt reasoned that this confluence could be attributed at least in part to the role that the natural sciences had played in dispelling "the special prestige of religious, hereditary, and state authority."[27] Putting the case against religion in especially strong terms, Levitt argued that "religion had to be annulled and diluted as doctrine . . . for either democratic politics or science to thrive."[28] Levitt was careful to qualify that science was not the "principle" or even the "sole" spark for the rise of liberal democracies, but he averred that "natural science has clearly been [a] source and model" for democratic flourishing.[29]

Although the views of those whom leftist critics called "science warriors" or advocates for the "new scientism" were not uniform, Levitt's musings on science and democracy distilled some essential points of the perspective of scientific liberals. In his view, a thorough acceptance of science dispelled superstitious and ideological reasoning. At its most extreme, scientific liberalism proposes that societies must be freed from the dogmatism of religion to achieve the rationalism needed for the civic contracts of democracy. To Levitt and other scientific liberals, the goal was not to make science itself democratic; in fact, his conception of science as something necessarily created through a merit-based elitism could hardly accommodate this possibility. Rather, scientific liberals contended that the egalitarianism of democracy depended on the elite expertise of science.

Levitt and Gross's *Higher Superstition* is generally considered the opening salvo of the science wars because of its vitriol against critical studies, particularly feminist studies. But it was an editorial review of the book that turned the science wars into a media sensation through what came to be known as the Sokal affair or the Sokal hoax. *Social Text*, an editor-curated journal published by Duke University, convened a special issue on the science wars for its spring/summer 1996 issue. Conceived specifically as a response to Levitt and Gross's *Higher Superstition*, the special issue gathered essays from notable figures from both the humanities and the natural sciences who had participated in the critique of science. Ruth Hubbard contributed a piece on the interplay between social gender roles and scientific definitions of biological sex, calling for "physicians and scientists . . . to remove their binary spectacles" and arguing that "major scientific distortions have resulted from ignoring similarities and overlaps in the effort to group differences by sex or gender."[30] Richard Levins, a noted population geneticist and philosopher, wrote a piece on the relationship between science and antiscience.[31] And Sarah Franklin, a feminist theorist, wrote about the cultural politics of the science wars.[32] The special issue was later republished as an edited volume with additional contributors; among them, notably, was Richard Lewontin.[33]

Significantly, the edited volume offered no reflection on the political activism of many of its contributors, several of whom had been key figures in the radical science movement, including publishing pieces in the magazine *Science for the People*.[34] Both Levins and Lewontin were prominent leftists and had coauthored a book on dialectical biology; as we have seen, Hubbard was an early leader of the Genes and Gender Collective.[35] But even if the influence of the radical science movement was only implicit, the volume still echoed the political and epistemological origins of the leftist view of science in the radical science movement.

It was the last piece in the *Social Text* special issue that provoked national and international media coverage. The article "Transgressing the Boundaries: Toward a Transformative Hermeneutics of Quantum Gravity" was later revealed to be an intentional parody of postmodern critiques of science by its author, the physicist Alan D. Sokal.[36] After revealing his scheme, Sokal claimed that the publication of his piece in the journal demonstrated the poverty of postmodern epistemology, since the editors had not recognized his piece as fake and had not sent it to any professional physicists for peer review. Bruce Robbins and Andrew Ross, the coeditors of *Social Text*, largely dismissed Sokal's actions as a "hoax."[37] In the ensuing years the academy was divided (primarily along disciplinary lines) over whether the affair had been a damning revelation of the lack of rigor in critical theory scholarship (including science studies) or a bad-faith act of academic fraud.

In different ways, both the science warriors and academic leftists were responding to growing public skepticism about scientific authority. While Levitt and Gross laid the blame squarely at the feet of postmodernism, Ross argued in the introduction to the *Social Text* special issue that the "rise of technoskepticism" in fact had come from widespread public recognition of the "allegiances of many of the sciences to elite interests—military, corporate, and state."[38] In Ross's reasoning, skepticism about or critiques of science did not stem from adherence to a particular political ideology. Rather, they were the consequences of widespread public recognition that science was complicit in nondemocratic institutions and practices. In striking contrast to Levitt's depiction of science as the individual work of meritocratic exceptionalism, Ross characterized science as a largely corporate and institutional endeavor. Ross took aim at the science warriors' attempts to "reinforce the myth of scientists as a beleaguered and isolated minority of truth-seekers, armed only with objective reasoning and standing firm against a tide of superstition." Arguing that the view of science as a "sequestered craftlike pursuit has been undermined by the wholesale proletarianization of scientific labor in commercial production," he professed that most scientists were in fact "industrial workers."[39]

Ross's most significant point had to do with the purpose and limits of critiquing science. He argued that "simply altering the socially constructed nature of the scientist's knowledge" was not enough to correct science's abuse of power or its constructed hierarchies. Rather, the actual practice of science required "methodological reform—to involve the local experience of users in the research process from the outset." It was only by doing this that "we begin to talk about different ways of doing science, ways that downgrade methodology, experiment, and manufacturing in favor of local environments, cultural values and principles of social justice."[40] This view that the practice of science itself must be more democratic for science to support social democracy underlay the critique of science during and after the science wars.

NOMA vs. Consilience

In a 2000 essay for *Science* titled "Deconstructing the 'Science Wars' by Reconstructing an Old Mold," Gould dismissed the debates of the science wars as merely the latest in a long history of "dichotomies" that had included the "17th-century debate of 'moderns'. . . versus 'ancients,'" "science versus religion in a favorite trope of late 19th-century secularism," and the "sciences versus the humanities in the icon for the second half of the 20th century." Gould considered the opposition between the "realists (including nearly all working scientists)" and the "relativists (nearly all housed in faculties of the humanities and social sciences)" to be merely the latest in a long line of "dichotomies [that] must be exposed as deeply and doubly fallacious."[41] Gould suggested instead that science and the postmodern humanities were not diametrically opposed but rather different ways of seeing the world that could peaceably coexist in their own spheres of influence.

Gould's article on the science wars echoed the pleas he had made a year earlier in *Rocks of Ages* when he proposed that NOMA was a framework that could support mutual goodwill between science and religion. Most of Gould's readers, including public scientists and theologians, understood NOMA precisely in the manner he presented it: as a timeless, ahistorical statement about the nature of science and religion. But the universality of NOMA's rhetoric belied its dependence on the immediate cultural politics of late– and post–Cold War America. Many reviewers correctly assumed that the ongoing creationist controversy had shaped Gould's thoughts on science and religion. Discussions of creationism dominated the book's reception, as many critics interpreted Gould's stance as a conciliatory position toward the movement.[42] But while creationism had indelibly transformed the rhetorical geography of American cultural politics, it was neither the only nor even the most important influence on Gould's thinking. A key component of the NOMA arguments was Gould's assertion that "nature just is . . . in all her sublime indifference to our desires . . . we cannot use nature for our moral instruction."[43] Since the earliest days of his public career, Gould had insisted that nature's amorality meant that humanity would have to define purpose, meaning, and ethics for itself. He consistently used and cited the humanities—particularly philosophy, history, and rhetoric—as a better guide than science for these aspects of human existence. NOMA was as much a plea for peace between the sciences and the humanities as it was a model for the proper relationship between science and religion.

By the time *Rocks of Ages* was published, Gould and Wilson's conflicts in the sociobiology controversy were several decades old. But even as the specific clash over sociobiology had cooled by the 1990s, their competing visions for the role of science in ethical matters and the proper relationship between science and humanistic critique had not faded away. One of the clearest abstractions of the differences between Wilson and Gould came from a pair of books they published at the close of the decade. Wilson released *Consilience: The Unity of Knowledge* in 1998, and Gould's response, titled *The Hedgehog, the Fox, and Magister's Pox: Mending the Gap Between the Sciences and the Humanities*, was published posthumously in 2003 In *Consilience*, Wilson expanded on his vision for the relationship between the sciences and the humanities, which he had first hinted at in his sociobiology writings during the 1970s. Recall that in 1976, in *Sociobiology*, Wilson had asked his readers to imagine the world "in the free spirit of natural history" so that they might see that "from this macroscopic view, the humanities and social sciences shrink to specialized branches of biology; history, biography, and fiction are the research protocols of human ethology."[44] Even as early as 1977, then, Wilson had asserted that there was no need for a respectful dialogue between the humanities and the sciences. He subsumed the humanities under the objectivity of the natural sciences.

In *Consilience*, Wilson gave this perspective a fuller treatment. Aptly subtitled *The Unity of Knowledge*, the book was Wilson's program for the "linkage of the sciences and the humanities." Wilson argued that the "Enlightenment thinkers of the seventeenth and eighteenth centuries got it mostly right the first time." In other words, the Enlightenment vision of a "lawful material world, the intrinsic unity of knowledge, and the potential of indefinite human progress" was not a dream that had to be set aside because of the criticisms of postmodernity and identity politics. The thoroughgoing triumph of the Enlightenment was, in fact, "increasingly favored by objective evidence . . . from the natural sciences." In Wilson's view, the fragmentation of political and philosophical truth by feminist, queer, and postcolonial theory did not reflect reality but was instead an "[artifact] of scholarship."[45] Furthermore, these myriad perspectives "[demote] the idea of a unifying natural culture."[46]

Throughout his writings, Gould tended toward a respectful approach to religion that recognized that science did not hold a monopoly on morals and ethics. By contrast, Wilson took a more aggressive stance, as in a chapter of *Consilience* titled "Religion and Ethics," where he argued that the emergence of modern science had given humanity reason to believe in an objective moral system. Wilson believed that evolutionary biology had taught us that moral sentiments were, in fact, evolved moral instincts. Thus the "ought" of moral questions was "the product of a material process," and thus there was a way forward to "an objective grasp of the origin of ethics."[47] Wilson collapsed what he took to be the evolutionary origin of morality with morality itself; knowing everything about the former, he asserted, would best prepare humanity to execute the latter. To Wilson, the natural sciences were the best foundation for an ethic of justice, freedom, and democracy.

Gould's response to Wilson's arguments came in *The Hedgehog, the Fox, and the Magister's Pox,* which was billed as a "spirited rebuttal to the ideas of [Gould's] intellectual rival E. O. Wilson."[48] In the book, Gould constructed a complicated analogy for his own view of the appropriate relationship between the sciences and the humanities. Gould borrowed his title from the philosopher Isaiah Berlin's famous essay "The Hedgehog and the Fox," which itself drew from both an ancient Greek proverb and the Renaissance writer Erasmus Roterodamus. In his essay, Berlin somewhat whimsically suggested that all great thinkers could be divided into two categories: "hedgehogs," who understand the world through one single defining idea, and "foxes," who draw on a wide variety of ideas to make sense of the world. In his book, Gould transposed this categorization of thinkers into a general principle for intellectual life. He suggested to his readers that they must be both foxes and hedgehogs in their approach to the sciences and the humanities. On the one hand, "these two great endeavors of our soul and intellect work in different ways and cannot be morphed into one simple coherence," so we must all be "foxes" in our intellectual life. On the other hand, "these two enterprises can lead us onward" to the "union of natural knowledge and creative art"—Gould's hope for the fullness of human existence.[49]

Gould's proposal for the relationship between the sciences and the humanities was nearly identical to his proposal for the relationship between science and religion. The fox-and-hedgehog analogy was intended to encourage "peace and mutual growth in strength," whereas NOMA was intended to assist those "people of goodwill who wish to see science and religion at peace, working together to enrich our practical and ethical lives."[50] In *The Hedgehog, the Fox, and the Magister's Pox,* Gould even referred to the "magisteria" of the sciences and the humanities.[51] Many of the examples and anecdotes he used were drawn from the same material as *Rocks of Ages.* The commonality between the two books merits the conclusion that Gould defended religion insofar as he saw it as one expression of humanistic knowledge.

Gould had first introduced his readers to NOMA in a *Natural History* column in the spring of 1997.[52] The opening anecdote recalled a trip almost fifteen years prior during which Gould "spent several nights at the Vatican housed in a hotel built for itinerant priests." He explained that he had been "in Rome for a meeting on nuclear winter sponsored by the Pontifical Academy of Sciences" and just happened to share the hotel with a group of "French and Italian Jesuit priests who were professional scientists." Gould relayed the experience to his readers in order to describe the Jesuits' puzzlement over the creationist controversy in the United States. In Gould's story, the Jesuits were worried that the controversy might mean that evolution was "really in some kind of trouble," which unsettled them, as they had, in their words, "always been taught that no doctrinal conflict exists between evolution and Catholic faith, and the evidence for evolution seems both entirely satisfactory and utterly overwhelming."[53]

This story accomplished several rhetorical objectives for Gould. It elevated his stature on the subject of science and religion by placing him in the company of the pope and the Pontifical Academy of Sciences at the Vatican. It allowed him to dismiss creationism as "a homegrown phenomenon of American sociocultural history—a splinter movement . . . of Protestant fundamentalists who believe that every word of the Bible must be literally true."[54] Finally, it presented him as the spokesperson for the magisterium of science, as he simultaneously argued that religion could "no longer dictate the nature of factual conclusions" and that neither could scientists "claim higher insight into moral truth from any superior knowledge of the world's empirical constitution."[55]

The immediate context for Gould's article, and the reason for his NOMA proposition, was a statement issued by Pope John Paul II the year before in an address titled "Truth Cannot Contradict Truth."[56] In it, the pope expanded on the Vatican's position on evolution, a position set out by his predecessor, Pope Pius XII, in 1950. In an encyclical titled *Humani generis,* Pius had affirmed the veracity of evolution with a tone of caution, regarding it as a serious hypothesis worthy of study.[57] In Gould's summary of the encyclical, Pius offered some "(holy) fatherly advice to scientists about the status of evolution as a scientific concept." Since the ideas were "not yet proven," Gould believed Pius was right to warn scientists to be cautious, since "evolution raises many troubling issues right on the border of [religion's] magisterium."[58] However, the intervening forty years had brought such a wealth of evidence in support of evolution that John Paul was keen to revise his predecessor's stance and declare that the theory of evolution stood on solid scientific ground.

It was in this essay that Gould used NOMA as a framework for the relationship between science and religion for the first time, but it was not the first time that Gould invoked the Vatican in a narrative about science and the social moral order. Although in 1997 Gould only briefly mentioned the reason for his stay at the Vatican in 1984, he had expended some amount of public energy in the years leading up to that visit alongside Carl Sagan as part of their effort to combat the expanding nuclear ambitions of the Reagan administration.[59] In defiance of a decades-long détente approach to the nuclear capacity of the Soviet Union, Reagan's administration made a series of moves in the early 1980s that signaling its belief that the United States could triumph in an exchange of nuclear weapons. Alongside antinuclear activists, Sagan and other scientists developed the theory of nuclear winter, which contended that even a small nuclear weapons event would release enough ash into Earth's atmosphere to cause climatic cooling and complete planetary devastation. During the nuclear winter debate, Sagan, Gould, and other scientists joined the public calls on the Vatican to combat the military stance of the US president. It was in this context that Gould first argued that religion could offer moral guidance that science was incapable of offering. Since the early days of his column, Gould had claimed that scientists' knowledge of the empirical world conveyed upon them no special moral authority. In the midst of the nuclear winter episode, Gould presented the other side of a developing argument: religion could provide the partnership with science. He could welcome "the support of a primary leader from the other major magisterium of our complex lives."[60]

The immediate reaction in 1999 to Gould's NOMA proposition was polarized. Many of Gould's readers were surprised when the long-standing critic of American creationism advocated for a religion to have a place equal to that of science in "the fullness of life." While some observers celebrated him for "illuminating the present with the past . . . and effectively cementing the principle of NOMA in the reader's mind by sheer force of example," others felt that *Rocks of Ages* was nothing more than a "badly written, condescending and misleading book."[61] Dawkins speculated about Gould's motivation, hypothesizing that the book was merely a political ploy meant to appease "those sensible religious people—the theologians, the bishops, the clergymen who believe in evolution."[62] Wilson concluded simply that Gould had "dodged the question."[63]

It was only in an unpublished essay titled "Why I Wrote *Rocks of Ages*" that Gould connected this discussion of science and religion with his vision of what it meant to be an American. He began the piece with his ideal vision of American democracy: "America, this wonderful country of ours, is both a bundle of contradictions and the greatest experiment in democracy ever attempted. . . . And yet, despite a range of threats from civil wars to school bombings, democracy has prevailed so far—primarily because Americans have practiced the art of respect, or at least toleration, for differences."[64] Gould's view of America was in keeping with his political and intellectual background. The grandson of recent immigrants from Eastern Europe, a New Left student protestor brought up by an activist father, his America was not one fashioned by a particular creed, ethnicity, or religious tradition. His vision of America had been formed simply by the ethos of tolerance and diversity that held its various demographics together. It was a resounding vision, and a relatively recent one.[65] In Gould's view, anything that threatened the country's tolerance for differences, including the kind of scientism advocated by some evolutionary biologists, was a threat to this American vision. He wrote, "Among the false divisions that threaten such tranquility the supposed warfare between science and religion ranks high among ideological sources of unnecessary strife."[66] Thus, in proposing NOMA Gould was launching an urgent appeal to other professional scientists writing and working in the American context. With NOMA, Gould sought a defense of humanistic knowledge that was a scientific insider's cautionary tale, an orthodoxy of appropriate scientific morality. Gould believed that scientists might best cultivate respectful tolerance in the American setting by curbing the overreaches of their own intellectual and social authority. At its best, science could "illuminate" but "never resolve" the realm of "human meaning, purposes and values."[67]

Thus, Gould's response to the science wars echoed his contention in *Rocks of Ages* that people of goodwill could find common ground if they recognized that their divisions were built on false dichotomies and overblown rhetoric. Gould's arguments about science and religion received far more press than did his views on the sciences and the humanities, in part because of the sensationalism of the creationist controversy and in part because the work that articulated this view most clearly, *The Hedgehog, the Fox, and the Magister's Pox,* was published posthumously. Nevertheless, he argued the same point on both fronts: that in a fractured world of battling ontologies, science had to cede its authority on morality and ethics to another cultural sphere. Whether this sphere called itself "the humanities" or "religion" made little difference to Gould. Scholars in the humanities must acknowledge the technical authority of science on empirical matters; scientists must also recognize the biases that inevitably arose from their social context.

Scholars in critical studies did not respond to Gould's propositions about science and the humanities with the same degree of criticism as scientists did to NOMA, but this should not necessarily be read as approval. Rather, they largely ignored it.[68] Gould contended that if science relinquished all pretenses to moral superiority, it would retain a social authority based upon its technical prowess. But these interdisciplinary scholars had a more radical vision, one of a future that rejected the arrangements of power and authority in Western societies. This future would not be achieved through the humble self-reflection of a few privileged white male scientists. The academic Left wanted to undo the heart of Western oppression, and if that required challenging the very foundations of modern science, they were willing to do so. Gould may have made his propositions in good faith, but they did little to bridge the chasm between liberal scientists and the critical humanities at the close of the twentieth century.

A New Atheism

After the events of September 11, the New Atheists catalyzed a version of scientific liberalism that revealed the ethnonationalist subtext of their definitions of Western science. The writings of Dawkins and Harris, in particular, racialized and sexed religious faith as weak, effeminate, and Brown. It also obscured the newness of religious expressions of modernity in both the United States and the Muslim world. Flood geology and Salafism were not relics of an ancient past; they were only conceivable as responses to the liberalism of nineteenth-century religious movements. But for New Atheists such nuances of history were only politically correct niceties that masked the true ignorance of religious faith. They took it upon themselves to defend Darwinian science as the crown jewel of enlightened rationality against the forces of dogma, superstition, and terrorism.

Two themes in particular stood out in these authors' writings. First, even as they asserted that Islam presented the most urgent threat to Western reason and liberty, they warned against all religion and supernatural belief. Harris made this vibrantly clear in *The End of Faith*, a book that characterized Islam as a moral and existential threat to the West. Harris exhorted his readers to recognize that "we are at war with Islam . . . with precisely the vision of life that is prescribed to all Muslims in the Koran."[69] Nevertheless, Harris explained, all religion, no matter how seemingly tolerant or liberal, should be considered the enemy of reason. Dawkins echoed this perspective when he proclaimed in *The God Delusion*, "I am attacking God, all gods, anything and everything supernatural, wherever and whenever they have been or will be invented."[70] "Our enemy," Harris wrote, "is none other than faith itself."[71] The New Atheists were particularly derisive of religious liberals, who advocated for legal and cultural deference toward religious faith. In their view, the problem was not solely or even primarily with religious extremists. Rather, they contended that the events of September 11 had been a tragic but almost inevitable result of the irrationalism of any type of supernatural belief.

The four horsemen also claimed that religious faith generally, and Islam specifically, was a throwback to an earlier, less rational period of human history. Harris used particularly vivid language to make this point. In *The End of Faith*, he asserted that "in our opposition to the worldview of Islam, we confront a civilization with an arrested history."[72] This narrative drew on a way of imagining the course of Western history in which Western civilization had taken a series of progressive steps away from dogmatism toward open-minded rationalism. According to these authors, the rise of Western liberal democracy had been possible because of the historical transformations of the Reformation, the Enlightenment, and the Scientific Revolution. In their view, these historical epochs moved Christianity away from the authoritarianism of Catholicism and set the stage for scientific naturalism to emerge as a dominant worldview. According to the New Atheists, this process had not occurred in the Islamic context, where an older, religious fundamentalism still ruled.[73] In advancing this narrative, the New Atheists championed a cosmology with a paradox at its heart. Science was that which was universally true and transcendent of culture, but it was also a uniquely Western achievement that had come about through the specific cultural transformations and innovations of Western science.

The New Atheists' ultimate message was that all intellectual movements that threatened the authority of Western science were an impediment to modernity and therefore to the liberal values of free speech, democracy, and respect for the rule of law. Religion was a charismatic culprit of irrationality, but they also singled out the cultural relativism of the postmodern academic Left as part of a cultural war on rationalism. These views had originated in the altercations between liberals and leftists during the sociobiology debates, and they had been further shaped by scientists' public encounters with right-wing creationism after the early 1980s. After this, liberals shifted their focus from championing science to attacking competing ideologies, a transformation that can be traced clearly in Dawkins's publishing career. In 1976 Dawkins released *The Selfish Gene*, which inserted him into cultural debates over the use of sociobiological claims in explaining sex difference and gender roles. Five years after the McLean trial, in *The Blind Watchmaker*, the target of his ire shifted from feminists to proponents of creation science. In 2004, three years after September 11, Dawkins argued in *The God Delusion* not only that science was the best argument against religious supernaturalism but that religion was dangerous. The claim had become that the future of a free, democratic society depended on an unequivocal championship of science. This progression represented more than just the idiosyncrasies of Dawkins's interests over the course of his public career; it traced a trajectory from a diffused scientific liberalism during the sociobiology debate, to heightened opposition to humanistic thinking during the science wars, and finally to a full-throated rejection of religion after September 11. Those who articulated and amplified this liberal scientific worldview came to see religion as the greatest enemy to democracy, science as its best protector, and feminists and other academic leftists as aiding and abetting dangerous global antiscience tendencies.

This evolution of scientific liberalism also explains why New Atheism so often aligned with the national rhetoric of the increasingly militarized US security state. In the public work of New Atheism, the championship of reason over religion became a justification for the (perhaps temporary) suspension of civil liberties for the sake of Western security. Harris, for instance, argued that "the only thing that currently stands between us and the roiling ocean of Muslim unreason is a wall of tyranny and human rights abuses that we have helped to erect."[74] Dennett adopted this same narrative in *Breaking the Spell*, lamenting, "If only we could figure out some way today to break the spell that lures thousands of poor young Muslim boys into fanatical madrasahs where they are prepared for a life of murderous martyrdom instead of being taught about the modern world, about democracy and history and science."[75] In this logic, a society deemed incapable of democracy can and should be governed through undemocratic means. Harris, in particular, implied that democracy was something that a culture earned by demonstrating its capacity for reason; that is, democracy was a privilege, not a right. Conversely, across the academic Left a growing body of scholarship critiqued the alignment of liberalism, US militaristic nationalism, and late capitalism. Although their analyses were not necessarily focused on the question of science and democracy, these scholars argued that liberalism as a doctrine, even more so than right-wing nationalism, was largely to blame for the evils of the war on terror. This critique of liberalism and neoliberalism further widened the gap that had opened up between academic liberals and leftists during the science wars.

Critical theorists and scholars of the academic Left responded to September 11 not by championing religion (as opposed to science) but by criticizing the nationalism, militarism, and neocolonialist actions of the United States in Afghanistan, Iraq, and throughout Central Asia. Even if these scholars did not explicitly frame these debates as a cosmological war between science and religion, they implicitly engaged with these themes through their critiques of the racialization of Islam that was so prominent in liberal writing during this decade. One of the most notable queer theorists to respond to September 11 was Jasbir Puar, a professor of women's and gender studies at Rutgers University. In a 2007 book, *Terrorist Assemblages: Homonationlism in Queer Times*, Puar argued that seemingly liberal narratives had been incorporated into American nationalism, which then had framed Muslim countries as necessarily oppressive to democratic liberties. She focused particularly on liberalizing narratives of gay rights in the United States, which contended that Western democracy afforded civil rights to queer communities. Concurrently, these narratives claimed that Muslim cultures were inherently oppressive to queer life. Puar used the term *homonationalism* to critique the assimilationist underpinnings of this contention and to undercut the narrative's claims of global liberation. She described how "queer subjects" had been transformed in the United States from "being figures of death (i.e., the AIDS epidemic) to becoming tied to ideas of life and productivity (i.e., gay marriage and families)."[76] Moreover, she argued that this incorporation of queer life into homonationalism was a distinctly racial project that only allowed queer subjects to be incorporated into citizenship once they were seen as productive for the American nation. Puar's bold critique pushed against assumptions that US society, and Western society more broadly, was truly liberatory.

Other scholars focused their critiques on the limitations on basic civil liberties for Muslim citizens in the United States, particularly after the passage of the USA PATRIOT Act in the immediate wake of the attacks. In the following decade, these academics chronicled the transformation of the category Arab / Middle Eastern / Muslim within US racial discourse from a state of "invisibility" to one of highly surveilled, visible subjects.[77] This literature highlighted the melding of these identity terms wherein religion and race were conflated. Through this process, Islam came to represent both a worldview and a racial group in the US cultural lexicon. And as Minoo Moallem, a professor of gender and women's studies at the University of California, Berkeley, argued in 2002, "Islamic fundamentalism has become a generic signifier used constantly to single out the Muslim other, in its irrational, morally inferior, and barbaric masculinity and its passive, victimized, and submissive femininity."[78] During these decades, the category "Muslim" emerged in American discourse as an inevitably raced religious identity—at odds with both a Christian ethnonationalist definition of US citizenship and the liberal vision of America as the standard-bearer of Western enlightenment.

As for Gould, the events of September 11 struck him very deeply. After all, he had grown up in Queens, and the Twin Towers were "less than a mile" from his home in Manhattan.[79] In a piece for the *New York Times* on September 26, he detailed some of his family's efforts to collect and distribute supplies and protective gear to first responders.[80] The tragedy also intersected with his extended family's history in the United States in a coincidence that Gould described in an op-ed for the *Boston Globe*. September 11, 2001, was precisely "100 years to the day" after Gould's maternal grandfather, Joseph Arthur Rosenberg, arrived in the United States from Hungary. Gould deemed the confluence between his family's American beginnings and the attacks on the Twin Towers the "most eerie coincidence that I have ever viscerally experienced."[81]

The piece was later reprinted as the coda to *I Have Landed*, Gould's tenth and final collection of his *Natural History* writings. In the coda, Gould articulated a vision of the state of American democracy that did not entirely align with either the views of the academic Left or those expressed by scientific liberals. Importantly, he had begun his career as an interloper within the patrician intellectual world of Harvard University. His sense of disaffection with his surroundings was most clearly exhibited in his writings on his working-class and Jewish identity. He relied on descriptions of his boyhood in Queens and playing stickball on the street to reach out to audiences beyond the ivory tower.[82] In the preamble to *Rocks of Ages*, Gould invoked his background as a member of a New York Jewish family with immigrant grandparents.[83] This was a signaling device that emphasized his origins outside the establishment of a WASP-ish academy and conveyed a family heritage that placed him within the reach of both nineteenth-century race science and the Holocaust. It was not dissimilar to the practice adopted by various feminist scholars who explicitly discuss their personal background and perspectives to emphasize the contextual and contingent nature of knowledge production. In the case of these and other scholars in critical disciplines, those choices drew upon a rich theoretical literature.[84] For them, the politics of scholarly identity were intermingled with reflections on their experiences in a twentieth-century academic world that struggled to incorporate diverse bodies into the life of the mind.

Although Gould entered the academy in the wave of New Left energy, by the end of the century his take on science and democracy was no longer aligned with developments among the academic Left. Critically, Gould described his NOMA proposal for peace between science and religion and, by extension, between science and the humanities as naturally appealing to "people of good will."[85] Gould's admonishments against what he characterized as the intellectual extremism behind both sides of these conflicts was rooted in a deep faith in the essential harmony between Western science and Western humanism. Whether writing about the science wars or the conflict between science and religion, Gould argued for the fundamental soundness of the Western intellectual legacy. In his view, democracy would not be secured by championing science over the humanities or by a radical rejection of the Western canon. Rather, by celebrating the essential wholeness of Western learning, people of "good will" would come to recognize that democracy itself was on solid footing. He ended *I Have Landed* with a soaring endorsement of American democracy: "We have landed. Lady Liberty still lifts her lamp beside the golden door. And that door leads to the greatest, and largely successful, experiment in democracy ever attempted in human history, upheld by basic goodness across the broadest diversity of ethnicities, economies, geographies, languages, customs, and employments that the world has ever known as a single nation."[86]

But the appeal of Gould's position faded somewhat in the context of post–September 11 America. During his career, Gould had attempted to construct a public image of Darwin that could accommodate religious belief and an appreciation for the humanities and would dismantle racial hierarchies. He also intended his version of Darwin to prove his own importance to the history of evolutionary theory. But no matter what Gould's motivations were, his Darwin found little space in the cultural battles over creationism, and even less in the anxieties of the post–September 11 American security state. In 2006, four years after Gould's death, *Wired* magazine ran a cover story featuring Dawkins, Dennett, Harris, and Hitchens that described the battles between Gould and the other evolutionists as "fading into history."[87]

There is no straightforward relationship between the rhetoric of the New Atheists and the actions of the United States and its national allies over two decades of the war on terror. Certainly, the cultural narratives that justified military action in Afghanistan and Iraq and the ongoing surveillance of Muslim, Middle Eastern, and other minoritized communities originated within the evangelical politics of the Bush administration. And they continued through the years of the Obama and Trump presidencies. The racialized discourse of a civilizational conflict with the "Muslim Middle East" reflected a broadly resonant—and persistent—view of Christianity as the cultural foundation of the United States. And so the fact that the New Atheist depictions of Islam as backward and illiberal resonated with these conservative arguments, even as New Atheists rejected evangelical religion, reveals the racial logic that tied all these narrative threads together. The New Atheists contended that civil liberties could only flourish in a society that committed itself to scientific knowledge as the guarantor of truth. But throughout their writings, they revealed an understanding of science that was distinctly identitarian, even tautological. That is, science was Western, and the West was science. In this logic, scientific liberalism and civil liberties were the exclusive purview of the West, defined as the United States and its allies. And through this definition, the West was justified in any measure—no matter how violent, invasive, or racist—to protect those democratic freedoms.

A Democratic Science beyond Cosmological Unity

Historians have described American evangelicalism as an "apocalyptic" social movement.[88] They typically mean this in the literal sense that many evangelicals believe the end-time is approaching and thus interpret American and global politics in light of the New Testament book of Revelation. It is, however, also a more general claim that evangelicals await a future in which their worldview will come to fruition and that those who disagree with them will be forced to acknowledge the truth of the Christian gospel. This apocalyptic vision helps fuel a singular sense of evangelical mission: to convert as many as possible to belief before the end actually arrives. The fact of belief is so overwhelmingly important that it can override all other considerations, including those practices—kindness, humility, love—that are claimed to be at the heart of the Christian message. We might uncharitably call this hypocrisy. More generously, we might understand it as a belief that systematic cosmology produces its own good fruits. That is, it matters less whether people practice the gospel, so long as they believe in it.

The evolutionary writings associated with the New Atheists were also infused with a vision of an end-time, of a utopia in which religious superstition will have been eradicated and peaceful harmony achieved through scientific rationality. Like evangelicals, Dawkins, Harris, and other New Atheist writers also argued that cosmology bears its own good fruits. Their policing of worldviews became more important than the very qualities they held to be hallmarks of science: empiricism, critical inquiry, open-mindedness.

But it is the very power of that science that makes this vigilance dangerous. The technologies, systems of political governance, and tacit modes of social oppression enabled by our knowledge of the natural world are at once the most extraordinary and the most terrifying accomplishments of humanity. When we understand science as coming into being through history and culture, we make space to critique it meaningfully and carefully. We cannot afford to have cartoonish or idealized versions of science even in defense against creationism and other reactionary political agendas. This is more than a question of being right about the definition of science or religion; it is a deeply political, ethical, and moral conversation.

The academic Left first took on Enlightenment science in its effort to create a science that could be oriented toward justice rather than oppression. Because they were primarily concerned with the use of scientific authority in claims about biological determinism, these scholars viewed the denaturalization of scientific epistemology as a key strategy for democratizing science. Their attempts to demystify science and its institutions took on many forms: they disrupted the professional niceties of academic journals and conferences, held community teach-ins, and spoke to public audiences through popular science media. These activities were aimed not only at public participation in science but also at creating transparent accountability between professional scientists and a wider society. It was over this last point that Gould and Wilson most strongly disagreed during the early days of the sociobiology controversy. But as these intuitions made their way into the humanistic methods in a series of critical fields, the work shifted from transparency to more radical criticisms of science's claims to neutral objectivity or unimpeachable rationality. This history reminds us, then, that the postmodern critique of science was not a simple unwillingness to face the hard truths of reality. Rather, it was born of a political movement to imagine and enact a democratic science.

This also suggests to us that if critique is a necessary mode of democratizing science, it is by no means the only way forward. From the 1960s on, groups on the left both inside and outside the academy worked toward a science for the people in ways that were not especially focused on humanistic arguments about the nature of scientific knowledge. Much of the explicitly leftist political work—such as the efforts of the Young Lords, the Black Panther Party, and the Combahee River Collective—focused on health care. These groups did more than disrupt conventions of how medical knowledge was delivered; they reimagined how health was defined and made accessible for the most historically marginalized in their communities. As we continue the work of democratizing science, there are a range of avenues we can and must pursue: racial, gender, and class diversity in STEM careers; public science education; citizen science; social accountability for funding structures; and public oversight of research ethics. Science that is done in broad community collaboration and without traditional hierarchies of expertise is often more effective than traditional science in delivering on the promises of health and well-being that liberal science advocates claim for science. This is well documented in cases of racial diversity; for example, Black doulas significantly reduce rates of morbidity and mortality among Black women delivering in hospitals, and Black doctors improve health outcomes for Black newborns.[89] But no single strategy will be enough on its own to upend the basic inequality at the center of modern science.

Leftist critique remains an especially powerful tool because it demonstrates that representative diversity within the existing frameworks and institutions of Western science is not sufficient for liberation. More is required than implicit bias training in science and medical education. Critique helps us to recognize that the very features of Western science that make it an effective form of natural knowledge are the same features that were born of colonialism, slavery, and genocide. Critique reorients our understanding of Western enlightenment and of secular science so that we characterize them not as bold steps taken by intellectual visionaries but as systems organized to master the natural world and the humans that reside in it. And critique encourages humility, so that we fully appreciate that there are other kinds of truth than Western science and other forms of knowing that make human flourishing possible. It is a humility that carries with it complexities: Is vaccine hesitancy on the part of Black communities reckoning with the nation's history of medical racism different from hesitancy on the part of white ethnonationalists suspicious of the deep state? Is the rejection of evolutionary science by Native Americans different from the attacks on it by white evangelicals? In this history, I have endeavored to show that challenges to Enlightenment science in the United States over the past half century have ranged across the political spectrum. Where it is possible for us to have empathy with these challenges tells us much about our own instinctive affiliations—whether we perceive the rejection of science to be from a lack of control or from an entitlement to dominance.

Still, I am not sure that a democratic science is possible. Can knowledge that comes from the mastery of the natural world ever be gained without the exploitation of some people? Perhaps there will be times when we will have to say that it is better not to know, better not to be in control, because the sacrifices entailed are too great. There will certainly be times when we must confront the fact that our own securities, whether in the form of effective drug treatments, the regulation of clean water and air, or the ability to assess medical risks, have come to us out of violence. We must hope that there will be other moments when scientific work is as healing and wondrous as we would most want it to be. When we can follow online communities of Black birders.[90] Or when the centralized health care system and culture of caring for elders leads to high COVID-19 vaccination rates on Native American reservations.[91] Or when mathematical models of Tla'amin Nation stone fish traps are part of computer education for university students.[92] When we will be able to better live in the joy of science because we are honest about its history. But we get nowhere with a Panglossian conviction that the historical and ongoing violence of science is the best of all possible worlds. We get nowhere by arguing that modern science is worth all its exploitation and sacrifice because it brings a better life to some of us. The people, all the people, will get nowhere.

What I end up with is not a single program, method, or plan for a democratic science. There will be things we try that will not work—when we are confronted with leaky pipelines for women and racial outsiders in STEM careers, when grassroots science is messy and unorganized, when demographic data reify categories rather than break them down, or when equitable funding sources dry up and innovative research is halted. There will be things we do not understand, that stretch and confuse us, when what we do will not conform to any of our assumptions about what science is or ought to be. And there will be moments that terrify us, when the work of democracy becomes an opening for tyranny and obfuscation. But we must at least believe that a democratic science is a good thing, and that it is a good enough thing to keep trying for, as hard as it will often be. Because when we recognize that science is, first and foremost, power; when we acknowledge that power is intrinsic to its practices because it is a knowledge that orders, classifies, observes, experiments, and engineers; when we acknowledge that even if it is not the only power in the world, it is a formidable one—when we confront all this, we must also face that with power always comes the temptation to forget the least of these. And so it is my hope that a democratic science is possible.

Epilogue

Why We Can't Wait for a Democratic Science

In December 2020, *CNN* reported that nearly one-third of the nurses who had died in the United States during the first year of the COVID-19 pandemic were Filipino or Filipino American.[1] Although Filipinos make up only 4 percent of the nursing population in the United States, they have accounted for a disproportionate number of deaths during the pandemic. One possible reason for this is that Philippine-trained nursing staff are overrepresented (in comparison with US-trained nurses) in ICUs, nursing homes, and long-term-care facilities owing to the long-standing imperial relationship between the mainland United States and the Philippines. Filipino nurses immigrated to the United States after the mid-twentieth century in response to staffing shortages. But despite a shared language and medical training, the ethnic divide and Euro-American nurses' lack of labor solidarity meant that Filipinos were more likely to work in frontline medical contexts.[2] The wrenching death toll of the COVID-19 pandemic has exposed decades-old structures of inequality and marginalization for Filipino health workers in the United States and around the globe.[3]

I read the *CNN* news story during the first pandemic year, after listening to a similar story on *NPR* over breakfast as I held a newborn in my lap.[4] But what was to be done with these facts, as so much horror washed over us all through a long year of pandemic and election news? The pandemic exposed the ugliest of realities in the United States—desperate income inequality, dramatic racial disparities in access to health care and health outcomes, the conspiratorial politics of public health mistrust, and the fearmongering of antidemocratic attacks on the right to vote. It also demonstrated our capacity for scientific innovation, as novel therapies and vaccine technologies met the needs of the moment. In a world in which we watched a president gloat over his access to experimental drugs even as he attacked democracy and turned away in cruel indifference to the country's dead, the promise of science felt like a necessary salvation of calm rationality. I found myself wondering whether it was ethical to critique science when so much was needed from it, when we all looked to science to do the extraordinary work of giving us our lives back. As I pored over the minutiae of clinical studies and CDC recommendations and confronted the ongoing Trumpian falsehoods and attacks on public health workers and scientific experts, I thought over and over to myself that it was not the time.

But the news story about Filipino nurses stayed with me. It was a story that reminded me that although my scholarly training came from an Ivy League graduate school, my body owed at least part of its existence to another world. The stories of loss seeped deep into my bones as I remembered the living rooms and garages of my childhood in Southern California, where I would sit with my Filipina grandma and aunties. I recalled times when my older cousin would affectionately call me "halfie" because my mother was the only white person at our extended family gatherings. I thought of the women in my family who had come to the United States in the 1970s to work as hotel maids and nail technicians. They did hair and cleaned houses. I thought of my aunts working as nurses and nannies in Saudi Arabia and Scotland.[5] As with all diasporic people and children of immigrants I have ever met, my Filipino heritage is at once a vivid personal reality and a reminder that my life is the result of forces far beyond my own control.

I think it is because of my family history that what consistently stands out to me about Stephen Jay Gould is his writings about his grandparents, all of whom were Jewish immigrants from Eastern Europe. He brought them into his public writing at many points—when finishing the run of his *Natural History* column, when discussing his vision for the relationship between science and religion, or, most importantly, in his history of nineteenth-century race science.[6] He also composed a long genealogy of his family, which he filed away in his papers for his sons.[7] It was this mixture of the intensely personal and the public that struck me; in it, I read a self-awareness on the part of Gould that, although he was a luminary in the most rarefied of intellectual circles, hardly a generation back he would have been on the losing side of Euro-American race science. To be Jewish and from Hungary might have made him the object of study and not the practitioner of evolutionary theory. Because of his reflections on his family, I have read in Gould, perhaps too strongly, a generation's worth of struggle over whether and how to use the scholarly life for justice rather than for personal glory.

I do not know whether Gould resolved this struggle for himself. Certainly, much of his effort, especially later in his career, was directed at securing his own scientific legacy. Nevertheless, I believe that this awareness of being a racial and intellectual outsider can be a kind of gift.[8] For me at least, the reminder that my body has not always been respectable in the life of the mind makes it impossible for me to have faith that everything will work out if we simply trust in the authority of science. In a hierarchical science there will always be those who are forgotten, abused, and exploited, even as others gain knowledge and control over the natural world. And there will always be diasporic people who bear the harshest burdens of sickness and death during a global pandemic.

We can't wait for a democratic science.[9] There will be no ideal time in the future when the authority of science is assured and it would thus be appropriate to deal with its violent history, to finally address its emergence out of empire, slavery, and patriarchy. Democratizing science is a way forward to knowledge and institutions that are for the people rather than for power. But it is by no means an assured path. A democratic science ceases to be an institution that any of us can predict or control. It opens possibilities that might dismay or even horrify us because there is no logically necessary political aim to the undoing of scientific authority. Even in the narrow history I have related here, the communities that contested Enlightenment science did so with opposing goals. The feminist biologists of the radical science movement criticized sociobiology as part of the women's movement for gender equality, whereas evangelicals pushed creationism into public classrooms as part of their effort to remake American in the image of fundamentalist biblical Christianity—an image that required conservative patriarchy and heterosexual marriage. Putting science in the hands of the people means all the people, including those who reject vaccines and sneer at the realities of climate change. It includes those who argue that masks are a liberal agenda and that queer love is a violation of nature. Even with all this, it is still necessary, because democratizing science is about survival.

★ ★ ★ ★ ★ ★ ★ ★

The Philippines became part of the United States during the expansion of America's imperial ambitions in the late nineteenth century. From its founding, the hunger for land and resources had pushed the young nation across the continent, fueling the genocidal policies of the Jackson administration of the 1830s and the war with Mexico in the following decade. Through treaties, purchases, theft, state-sanctioned murder, and the attendant effects of disease and famine the United States took territory from Native tribes until it could claim that America stretched "from sea to shining sea." The ideals of Manifest Destiny were tied also to the country's original sin; a driving force in the push into western territories was the desire to open land to slave-owning cotton planters.[10]

Commercial interests also fueled the war with Spain that ended in 1898 with the United States in possession of the Philippines, Puerto Rico, and Guam. The creation of an overseas empire helped establish the United States as a regional power in the Pacific and the Caribbean and propelled it onto the global stage as a naval military force to be reckoned with. Imperialism and America's self-conception as a nation of unique liberty were partners in this endeavor. US politicians justified the war to domestic audiences on the pretext of saving Spain's colonized populations from the tyranny of Catholic authoritarianism.[11] Political cartoons of the era featured images of despondent Filipinos, Cubans, and Puerto Ricans waiting to be saved by America's Lady Liberty.[12] But this political rhetoric did not mean that the United Stated held to a principle of self-determination for the peoples it had conquered. As the war hero and later president Theodore Roosevelt argued in 1899, the United States "had a legacy of duty"; it could not drive out "medieval tyranny only to make room for savage anarchy . . . some stronger manlier power would have to step in to do the work, and we would have shown ourselves weaklings."[13] Despite movements for Filipino independence, the United States maintained control over the islands until the Japanese invasion during World War II, after which the Philippines gained independence in 1945.[14]

Losing part of its Pacific holdings did not mean that the United States ceased to be a colonial power in the latter part of the twentieth century. If anything, the 1947 Truman Doctrine, which declared that the US government would commit military and economic assistance to all democratic nations under threat from authoritarian force, set the country on a path of even greater effort to shape the course of the postcolonial states of the emerging Third World. In the face of its military and nuclear standoff with the communist Soviet Union, America exerted its military, economic, and bureaucratic might around the globe in a hitherto unseen capacity.[15] Historians and critical theorists continue to reckon with the role of empire in shaping America's sacred narratives and moral language.[16] And for those most deeply oppressed, there was no forgetting that the United States was an empire. In 1979 Barbara Smith and the other radical Black feminists of the Combahee River Collective declared themselves to be "Third World Women."[17] It was a statement of solidarity with the peoples whose states had emerged from the legacy of European colonialism and were now buffeted by the militarized interventions of NATO and the Warsaw Pact. That solidarity was not merely rhetorical. It was a felt reality that reflected their status as a population whose health and reproduction were managed by the biopolitical American state.

The expanding moral circle is central to the liberal narrative of American history. In this metaphor, the founding of the nation was a bold, if incomplete, step toward the achievements of universal suffrage and social equality. But when we reckon with America as an inherently colonial entity, we see that it is not a nation that is set and formed and then gradually expanded. In a postcolonial frame, the nation is an unsettled material creation, and the history of the United States is not that of an expanding moral circle but rather that of a physical, material, and geographic negotiation over what land the nation will cover, which populations will be exterminated, which will be bred, and which will be enfolded into limited forms of recognition or into full citizenship. It is a racial creation not just in the abstract or even jurisprudential definition of race categories but in the embodied, physical and reproductive enactment of state policies of immigration, family planning, environmental conservation, forcible resettlement, migration detention, incarceration, enslavement, and genocide.

The demographic transformations of the United States during the twentieth century were managed by the scientized governance of a biopolitical and neoliberal state. But the unsettling of the nation's racial character also fueled the ethnonationalism of the emerging Christian Right. When Timothy LaHaye and Henry Morris began the ICR in 1972, they built it in El Cajon, a suburb of San Diego that was home largely to working- and middle-class white families. Through the early part of the century, it had been a primarily agricultural area, known for growing grapes and citrus. Along with the rest of the region, it had been ceded to the United States by Mexico in the 1848 Treaty of Guadalupe Hidalgo. And although it retained that Mexican heritage, in the immediate period after the Mexican-American War non-Hispanic white groups moved to the area. Even more arrived during the western migrations of the post-Depression era and the economic expansion of the Sunbelt suburbs in the 1950s and 1960s. But soon after the ICR opened its doors, the area began experiencing a dramatic racial transformation as new immigration from Latin America and the Arab world brought new communities to the city. Hookah shops and Mexican tortilla shops replaced Thrifty's drugstores. Spanish and Arabic signs became more common than English ones.

I know all this from historical reading and research. But I also know it because my childhood home was less than five miles from the offices of the ICR. By the time I was a teenager, the El Cajon middle school where I was a student and my mother taught had a majority nonwhite student population, with more than half a dozen languages represented on campus. Many of the students had parents who worked on both sides of the Mexico border, and by the next decade many would come to school directly from the refugee camps of America's war on terror. This was the world in which the ICR published *Acts and Facts*, LaHaye preached from the pulpit of Shadow Mountain megachurch, and students attended Christian Heritage, the Bible college LaHaye founded on the same campus. Creationism may have had deep roots in America soil, but it was given a new life in this era of racial and religious change.

I also know this because I attended Skyline Church, a sister megachurch to Shadow Mountain, and many of my friends and youth group leaders were college students at Christian Heritage. Skyline's senior pastor, Jim Garlow, was a central figure in the evangelical attack on gay civil liberties. He was a leading force behind Proposition 8, a 2008 effort to amend the California state constitution to define marriage as being between one man and one woman. He would later serve on a pastoral advisory council to the Trump administration. Garlow baptized me into the faith when I was seventeen. I still have the photo of myself after the baptism, exultant and ready to carry those convictions with me into the wider world.

When I was in high school in the late 1990s, in the years before September 11, the primary cultural battleground for evangelicals was gender and sexuality. This was the era of purity balls and the rise of the Silver Ring Thing ministry.[18] The Council on Biblical Manhood and Womanhood developed a complementarian theology in its war with second-wave feminism.[19] And threaded through this fight against feminists and the sexual revolution was an apocalyptic fervor that looked always toward the end-times. I read LaHaye's *Left Behind* series—sixteen novels that depicted in gory detail the experience of the world after the faithful are raptured home to Jesus—in the same years as I studied *When God Writes Your Love Story* and *I Kissed Dating Goodbye.*[20] To be sexually pure, gender-conforming, and pro-life was to join the spiritual battle of the apocalypse.

But in the years after I left El Cajon, evangelicals' preoccupations shifted from sexual purity to ethnonationalism. The movement's underlying discomfort with non-European immigration shot to the surface in the wake of September 11. The ethnonationalism that had been subtext for the preceding decades erupted as vibrant text as evangelicals now saw America as a Christian nation in a holy war with the Muslim Middle East. After I left Skyline, Garlow's ministry expanded from railing against gay marriage to diatribes against Islam.

Reading Gould's hopeful vision for America's future in the last decade of his life alongside this history, it is hard not to conclude that it was a dream divorced from reality. His was not the dream of the evangelicals, who desired a national identity wedded to a racialized Christianity. Nor was it the dream of the scientific liberals, who viewed America as the civilizational apex of the Western enlightenment. He had a far more inclusive vision than either of these, a vision wrapped in the promise of Lady Liberty, in which persecuted immigrants would come to the United States and establish a multiracial and multifaith democracy. It was a dream that Gould nurtured in the Jewish immigrant communities in Queens in which he was raised. Growing up in the shadow of the Holocaust and with the memory of Europe's long history of religious wars, he imagined a robust form of tolerance in which all people of goodwill could come together in peace and respect. But Gould's dream did not reckon with a world in which America was not simply the kindly land that welcomed refugees but also the imperial power whose military and economic might created the need for other peoples to flee their homes. Or in which the United States had formed itself through the genocide of America's original peoples. And where slavery was not only its original but also its abiding sin as it produced new forms of racial control in the architecture of its prison and migrant detention centers.

How is such a history to be reckoned with? Moreover, how are we to contend with the scale at which scientific knowledge, institutions, and technologies are interwoven into this history? Any kind of optimism or hope in the face of this seems foolish at best.

And so, I conclude not with hope but with what I have learned from listening to the activists, scholars, and thinkers who believed that criticizing science could and must be part of the work of justice for the United States. From Gould I learned the confidence to reject Panglossian explanations of the world as it is—to refuse to believe that the violences of the world are logical necessities or adaptive imperatives. From feminist and critical race critics I learned that the benefits of science have come to us as a result of oppression and exclusion. And I learned from leftist health collectives the power of populist action and the necessity of solidarity among the peoples made into populations managed by American colonialism.

I went looking for the answer to what I thought was essentially an abstract question—could a democratic society rest upon the authority of science if that science was not itself a democratic form of knowledge? What I came to realize is that philosophical debates over the nature of science were always about the material work of creating the nation-state. And that these narratives were enfleshed as biopolitical realities and made sacred across evangelical texts, immigration policies, defenses of the Western enlightenment, and denouncements of Muslim religion. I don't believe that scientific liberalism can offer us what we need to reckon with the history of enslavement, patriarchy, and colonialism that built the United States. The defense of science cannot tell us who we would welcome, nor can it provide sanctuary. It cannot reveal to us the harms of scientific racism and sexism. It cannot cultivate the humility we need to hear the hard truths that may destabilize everything we know and love.

It is through holding ourselves more firmly to history, through opening ourselves to the radical possibilities engendered by the criticisms of people who live across borders and whose ancestors perished in enslavement and genocide, that justice may come. Perhaps it is religion that inspires us to relinquish our desire to horde resources and power. Perhaps it is activism that moves us to understand what reparations are and must be. Perhaps it is empiricism that makes clear the depth and breadth of harm by the state. A democratic science is one that is willing to sit in the chaos and confusion of a worldview that has no consilience, and it is open to an American democracy that is deliberately postnationalist and even postnation. It is a science that gives itself over to the people, a science that is willing to seem not like science, that is flexible in its epistemologies, gentle in its practices, and diasporic in its intuitions. This may be a democratic science that can help rather than harm, that can be and do for what we need it to be.

Introduction

1. In July 2020 the Penn Museum removed the Morton Collection from public view and formed a committee to evaluate steps to repatriate or bury the crania of enslaved persons in the collection. In January 2024, nineteen of the twenty Black Philadelphians whose remains were in the collection were buried at Eden Cemetery in Philadelphia. See https://www.penn.museum/about-collections/statements-and-policies/morton-cranial-collection, accessed April 2024.

2. Ann Fabian, *The Skull Collectors: Race, Science, and America's Unburied Dead* (Chicago: University of Chicago Press, 2010).

3. Myrna Perez Sheldon, "Breeding Mixed-Race Women for Profit and Pleasure," *American Quarterly* 71, no. 3 (2019): 741–65.

4. Fabian, *Skull Collectors*, 2–3; Terence Keel, *Divine Variations: How Christian Thought Became Racial Science* (Palo Alto: Stanford University Press, 2018), 5–77; James Poskett, *Materials of the Mind: Phrenology, Race, and the Global History of Science, 1815–1920* (Chicago: University of Chicago Press, 2019), 19–50.

5. University of Pennsylvania Museum of Archaeology and Anthropology, "Museum Announces the Repatriation of the Morton Collection," news release, 21 April 2021, https://penntoday.upenn.edu/news/penn-museum-announces-repatriation-morton-cranial-collection.

6. Jason E. Lewis, David DeGusta, Marc R. Meyer, Janet M. Monge, Alan E. Mann, and Ralph L. Holloway, "The Mismeasure of Science: Stephen Jay Gould versus Samuel George Morton on Skulls and Bias," *PLOS Biology* 9, no. 6 (2011).

7. Lewis et al., 1; Stephen Jay Gould, *The Mismeasure of Man* (New York: Norton, 1981).

8. Gould, *Mismeasure of Man*, 69, 30.

9. Audra J. Wolfe, *Freedom's Laboratory: The Cold War Struggle for the Soul of Science* (Baltimore: Johns Hopkins University Press, 2018), 210–21.

10. "*Mismeasure of Man*, research and notes," Box 263, Folder 4, Stephen Jay Gould Papers, M1437, Department of Special Collections, Stanford University Libraries, Stanford, CA (hereafter Gould Papers).

11. Michael Weisberg, "Remeasuring Man: Remeasuring Man," *Evolution & Development* 16, no. 3 (2014): 166–78; Michael Weisberg and Diane B. Paul, "Morton, Gould,

and Bias: A Comment on 'The Mismeasure of Science,'" *PLOS Biology* 14, no. 4 (2016): 2.

12. Jonathan Michael Kaplan, Massimo Pigliucci, and Joshua Alexander Banta, "Gould on Morton, Redux: What Can the Debate Reveal about the Limits of Data?," *Studies in History and Philosophy of Science Part C: Studies in History and Philosophy of Biological and Biomedical Sciences* 52 (2015): 22–31; Paul Wolff Mitchell, "The Fault in His Seeds: Lost Notes to the Case of Bias in Samuel George Morton's Cranial Race Science," *PLOS Biology* 16, no. 10 (2018): e2007008.

13. Fabian, *Skull Collectors*, 3–4.

14. Lewis et al., "Mismeasure of Science," 5.

15. Fabian, *Skull Collectors*, xv.

16. Sheldon, "Breeding Mixed-Race Women for Profit and Pleasure," 754–56.

17. Stephen Jay Gould, "Biological Potential vs. Biological Determinism," *Natural History* 85, no. 5 (1976): 22; Gould, *Mismeasure of Man*, 20–21.

18. Steven Pinker, "The Intellectual War on Science," *Chronicle of Higher Education*, 13 February 2018.

19. Joanna Radin, *Life on Ice: A History of New Uses for Cold Blood* (Chicago: University of Chicago Press, 2017); Myrna Perez Sheldon, "Race and Sexuality: A Secular Theory of Race," in *A Cultural History of Race in the Age of Empire and Nation State*, ed. Marina B. Mogilner (London: Bloomsbury Academic, 2021), 149–64; Rana A. Hogarth, *Medicalizing Blackness: Making Racial Difference in the Atlantic World, 1780–1840* (Chapel Hill: University of North Carolina Press, 2017); Deirdre Benia Cooper Owens, *Medical Bondage: Race, Gender, and the Origins of American Gynecology* (Athens: University of Georgia Press, 2017); Rebecca Skloot, *The Immortal Life of Henrietta Lacks* (New York: Crown, 2009).

20. Skloot, *Immortal Life of Henrietta Lacks*, 102–4, 212–15.

21. "Significant Research Advanced Enabled by HeLa Cells," National Institutes of Health, accessed April 2021, https://osp.od.nih.gov/scientific-sharing/hela-cells-landing/.

22. "About the Foundation," The Henrietta Lacks Foundation, accessed 1 May 2021, http://henriettalacksfoundation.org/about/.

23. Owens, *Medical Bondage*.

24. Gabriela Soto Laveaga, *Jungle Laboratories: Mexican Peasants, National Projects, and the Making of the Pill* (Durham, NC: Duke University Press, 2009); "Letter to William McDermott," 18 October 1960, folder 219, Mary Steichen Calderone Papers, 1904–1971, Schlesinger Library, Radcliffe Institute, Harvard University, Cambridge, MA.

25. Michiko Kakutani, *The Death of Truth: Notes on Falsehood in the Age of Trump* (New York: Crown, 2018); Steven Pinker, *Enlightenment Now: The Case for Reason, Science, Humanism, and Progress* (New York: Penguin Books, 2018).

26. Owens, *Medical Bondage*, 15–41; Brianna Theobald, *Reproduction on the Reservation: Pregnancy, Childbirth, and Colonialism in the Long Twentieth Century* (Chapel Hill: University of North Carolina Press, 2019), 147–72; Centers for Disease Control and Prevention, "COVID-19 in Racial and Ethnic Minority Groups," 25 June 2020.

27. Jamie Cohen-Cole, *The Open Mind: Cold War Politics and the Sciences of Human Nature* (Chicago: University of Chicago Press, 2014); Wolfe, *Freedom's Laboratory*.

28. James Gilbert, *Redeeming Culture: American Religion in an Age of Science* (Chicago: University of Chicago Press, 2008), 5; Everett Mendelsohn, "The Politics of Pessimism: Science and Technology circa 1968," in *Technology, Pessimism, and Postmodernism*, ed. Yaron Ezrahi, Everett Mendelsohn, and Howard Segal, *Sociology of the Sciences* 17 (Dordrecht, Netherlands: Springer Dordrecht, 1994), 151–73; Andrew Jewett, *Science, Democracy, and the American University: From the Civil War to the Cold War* (Cambridge: Cambridge University Press, 2012), 2.

29. Kelly Moore, *Disrupting Science: Social Movements, American Scientists, and the Politics of the Military, 1945–1975* (Princeton, NJ: Princeton University Press, 2009). For histories of science that have explored counterculture elements as inspirational to theoretical science in this period, see David Kaiser, *How the Hippies Saved Physics: Science, Counterculture, and the Quantum Revival* (New York: Norton, 2011); W. Patrick McCray, *The Visioneers: How a Group of Elite Scientists Pursued Space Colonies, Nano-technologies, and a Limitless Future* (Princeton, NJ: Princeton University Press, 2013); and Michael D. Gordin, *The Pseudoscience Wars: Immanuel Velikovsky and the Birth of the Modern Fringe* (Chicago: University of Chicago Press, 2012), 165–66.

30. On the military-industrial-academic complex, see Stuart W. Leslie, *The Cold War and American Science: The Military-Industrial-Academic Complex at MIT and Stanford* (New York: Columbia University Press, 1993). For the development of "cold war rationality," see Paul Erickson, Judy L. Klein, Lorraine Daston, Rebecca Lemov, Thomas Sturm, and Michael D. Gordin, *How Reason Almost Lost Its Mind: The Strange Career of Cold War Rationality* (Chicago: University of Chicago Press, 2013). For a general overview of Cold War science, see Naomi Oreskes and John Krige, *Science and Technology in the Global Cold War* (Cambridge, MA: MIT Press, 2014). The literature on the environmental movement of the 1960s and 1970s is vast. Scholarship on the science activist Rachel Carson has been particularly effective in pinpointing shifting views of science after the postwar period in relationship to rising concerns about pollution. See David K. Hecht, "Constructing a Scientist: Expert Authority and Public Images of Rachel Carson," *Historical Studies in the Natural Sciences* 41, no. 3 (2011): 277–302.

31. Alondra Nelson, *Body and Soul: The Black Panther Party and the Fight against Medical Discrimination* (Minneapolis: University of Minnesota Press, 2011), 23.

32. The *Radical Science Journal* was published from 1974 to 1985, and *Science for the People* ran from August 1970 through May/June 1989 and then restarted publication in 2019. For the organizational history of the movement see Kelly Moore, "Organizing Integrity: American Science and the Creation of Public Interest Organizations, 1955–1975," *American Journal of Sociology* 101, no. 6 (996); and Sigrid Schmalzer, Daniel S. Chard, and Alyssa Botelho, eds., *Science for the People: Documents from America's Movement of Radical Scientists* (Amherst: University of Massachusetts Press, 2018).

33. On the grassroots nature of the antinuclear movements, see Robert Gottlieb, *Forcing the Spring: The Transformation of the American Environmental Movement* (Washington, DC: Island Press, 2005), 235–43. For examples of self-identified feminist biologists, see Linda Marie Fedigan, *Primate Paradigms: Sex Roles and Social Bonds* (Chicago: University of Chicago Press, 1982); Donna J. Haraway, *Primate Visions: Gender, Race, and Nature in the World of Modern Science* (New York: Routledge, 1989);

and Linda Marie Fedigan, "The Paradox of Feminist Primatology: The Goddess's Discipline?," in *Feminism in Twentieth Century Science, Technology and Medicine*, ed. Angela Craeger, Elizabeth Lunbeck, and Londa Schiebinger (Chicago: University of Chicago Press, 2001).

34. The phrase *too important to be left to scientists* was often used in the UK-based science news weekly *New Scientist*. See, e.g., John Langrish, "Rothschild is Right for Industry," *New Scientist* 53, no. 779 (1972): 144; Patrick Fisher, "Science Policy and Trade Unions," *New Scientist* 55, no. 303 (1972): 17; and John McCreesh, "Why Bother about Science?," *New Scientist* 57, no. 837 (1973): 612.

35. Gould, *Mismeasure of Man*, 21–22.

36. The term *scientific creationism* came to prominence in creationist circles primarily through Henry M. Morris's *Scientific Creationism* (Green Forest, AR: New Leaf Publishing Group, 1974).

37. Dorothy Nelkin, *Science Textbook Controversies and the Politics of Equal Time* (Cambridge, MA: MIT Press, 1977), 41–117.

38. See, e.g., chapter 4, "Uniformitarianism and the Flood: A Study of Attempted Harmonizations," and chapter 5, "Modern Geology and the Deluge," in *The Genesis Flood: The Biblical Record and Its Scientific Implications*, by Henry M. Morris and John Clement Whitcomb (Philadelphia: Presbyterian and Reformed Publishing, 1961), 89–115 and 116–211.

39. See, e.g., the introduction to the second edition of Henry M. Morris, *Scientific Creationism* (Green Forrest, AR: New Leaf Publishing Group, 1985), vii.

40. Ronald L. Numbers, *The Creationists: From Scientific Creationism to Intelligent Design*, 2nd ed. (Berkeley: University of California Press, 2006), 312–28; Jeffrey P. Moran, *American Genesis: The Antievolution Controversies from Scopes to Creation Science* (New York: Oxford University Press, 2012), 20–24; Susan L. Trollinger and William Vance Trollinger Jr., *Righting America at the Creation Museum*, Medicine, Science, and Religion in Historical Context (Baltimore: Johns Hopkins University Press, 2016).

41. For a consideration of reactions to the Christian Right among leftist intellectuals, see Daniel T. Rodgers, *Age of Fracture* (Cambridge, MA: Harvard University Press, 2011), 165–78; and Andrew Hartman, *A War for the Soul of America: A History of the Culture Wars* (Chicago: University of Chicago Press, 2015), 70–101.

42. A number of historians locate the origins of the Christian Right in the "culture wars" of the 1970s, pointing to various issues as the source for this new political entity. See, e.g., William Martin, *With God on Our Side: The Rise of the Religious Right in America* (New York: Broadway Books, 1996); Kenneth J. Heineman, *God Is a Conservative: Religion, Politics, and Morality in Christian America* (New York: New York University Press, 1998); and Ruth Murray Brown, *For a "Christian America": A History of the Religious Right* (Amherst, NY: Prometheus Books, 2002). Daniel K. Williams argues that the 1980s did not represent a new desire for political action on the part of evangelicals but rather a new tactic of political engagement that located evangelical efforts exclusively within the structures of the Republican Party. See Williams, *God's Own Party: The Making of the Christian Right* (Oxford: Oxford University Press, 2012), 2.

43. A notable example of accusations that the academic Left was aligned with creationism is in Paul R. Gross and Norman Levitt, *Higher Superstition: The Academic Left and Its Quarrels with Science*, 2nd ed. (Baltimore: Johns Hopkins University Press,

1998), 3, 6. The field of feminist science studies in particular has been accused of being "anti-realist" or "anti-science." For reflections and rebuttals of this association, see Sarah S. Richardson, "Feminist Philosophy of Science: History, Contributions, and Challenges," *Synthese: An International Journal for Epistemology, Methodology and Philosophy of Science* 177, no. 3 (2010): 337–62; and Elisabeth Lloyd, "Science and Anti-Science: Objectivity and Its Real Enemies," in *Feminism, Science and the Philosophy of Science*, ed. L. H. Nelson and J. Nelson (Boston: Kluwer Academic, 1996).

44. Ava Kofman, "Bruno Latour, the Post-Truth Philosopher, Mounts a Defense of Science," *New York Times Magazine*, 25 October 2018.

45. Kofman.

46. Bruno Latour, "Why Has Critique Run out of Steam? From Matters of Fact to Matters of Concern," *Critical Inquiry* 30, no. 2 (2004): 230.

47. Latour, 232.

48. Pinker, *Enlightenment Now*, 6.

49. Evolutionary science comprises far more than human evolution, but humans are the focus of the controversies discussed in this book.

50. Here I use *circulated* in the sense that James Secord advocates in his argument that scientific knowledge is made in the process of movement, translation, and transmission. It is expressly posited against dissemination as a model of the relationship between professional and popular science. Secord, "Knowledge in Transit," *Isis* 95, no. 4 (2004): 654–72.

51. Richard Dawkins, *Selfish Gene* (New York: Oxford University Press, 1976), 2.

52. Robert Ardrey, *African Genesis: A Personal Investigation into the Animal Origins and Nature of Man* (New York: Atheneum, 1961); Ardrey, *The Hunting Hypothesis: A Personal Conclusion Concerning the Evolutionary Nature of Man* (New York: Atheneum, 1976); Richard E. Leakey and Roger Lewin, *Origins: What New Discoveries Reveal about the Emergence of Our Species and Its Possible Future* (New York: Dutton, 1977). For a historical discussion of the ancestry of man, see Nadine Weidman, "Popularizing the Ancestry of Man: Robert Ardrey and the Killer Instinct," *Isis* 102, no. 2 (2011): 269–99.

53. Edward O. Wilson, *Sociobiology: The New Synthesis* (Cambridge, MA: Belknap Press of Harvard University Press, 1975), 553.

54. Sarah Blaffer Hrdy, "Empathy, Polyandry, and the Myth of the Coy Female," in *Feminist Approaches to Science*, ed. Ruth Bleier (New York: Pergamon, 1986).

55. Stephen Jay Gould, "The Evolutionary Biology of Constraint," *Daedalus* 109 (1980): 41–42.

56. Gould, *Mismeasure of Man*, 24.

57. Stephen Jay Gould and Richard C. Lewontin, "The Spandrels of San Marco and the Panglossian Paradigm: A Critique of the Adaptationist Programme," *Proceedings of the Royal Society of London. Series B. Biological Sciences* 205, no. 1161 (1979): 581–98. The paper is discussed at length in chapter 3. For an overview of the responses over two decades, see Massimo Pigliucci and Jonathan Kaplan, "The Fall and Rise of Dr Pangloss: Adaptationism and the Spandrels Paper 20 Years Later," *Trends in Ecology & Evolution* 15, no. 2 (2000): 66–70.

58. Gould began to promote this hierarchical theory in two papers he published in 1980, which then formed the basis for his scientific agenda for the ensuing decade. Stephen Jay Gould, "The Promise of Paleobiology as a Nomothetic, Evolutionary

Discipline," *Paleobiology* 6, no. 1 (1980): 96–118; Gould, "Is a New and General Theory of Evolution Emerging?," *Paleobiology* 6, no. 1 (1980): 119–30. For a complete discussion of this period in Gould's career and its relationship to the emergence of paleobiology, see David Sepkoski, *Rereading the Fossil Record: The Growth of Paleobiology as an Evolutionary Discipline* (Chicago: University of Chicago Press, 2012), 371–82.

59. Stephen Jay Gould, *Wonderful Life: The Burgess Shale and the Nature of History* (New York: Norton, 1989), 35.

60. Gould, 45.

61. Stephen Jay Gould, "Darwin's Dilemma," *Natural History* 83, no. 6 (1974): 24–29, republished as "Darwin's Dilemma: An Odyssey in Evolution," in Gould, *Ever Since Darwin: Reflections in Natural History* (New York: Norton, 1977), 38.

62. Gould, *Mismeasure of Man*, 38–39.

63. Elisabeth Anne Lloyd, *The Case of the Female Orgasm: Bias in the Science of Evolution* (Cambridge, MA: Harvard University Press, 2005). Lloyd's critique of an adaptive explanation for the female orgasm is discussed further in chapter 5.

64. The connection between evolution, materialism, and moral relativism was explicit in the writings of both Jerry Falwell and Timothy LaHaye. See Jerry Falwell, *Listen, America!* (New York: Doubleday, 1980); and Tim LaHaye, *The Battle for the Mind*, Power Books (Old Tappan, NJ: Revell, 1980).

65. Myrna Perez Sheldon, "Claiming Darwin: Stephen Jay Gould in Contests over Evolutionary Orthodoxy and Public Perception, 1977–2002," *Studies in History and Philosophy of Science Part C: Studies in History and Philosophy of Biological and Biomedical Sciences* 45 (March 2014): 139–47.

66. Steven Pinker, "Science Is Not Your Enemy," *New Republic*, 6 August 2013, https://newrepublic.com/article/114127/science-not-enemy-humanities; Mark Lila, "The End of Identity Liberalism," *New York Times*, 18 November 2016; Sam Harris, "Forbidden Knowledge: A Conversation with Charles Murray," 22 April 2017, *Making Sense* (podcast); Pinker, *Enlightenment Now*, 29–36.

Chapter 1 • Racism and the Promise of a Democratic Science

Epigraph: Stephen Jay Gould, "Racist Arguments and I.Q.," *Natural History* 83, no. 5 (June 1974): 29. All epigraph sources are by Gould.

1. "Ohio Barber Shop Shuts in Protests," *New York Times*, 16 March 1964.

2. Literature kept by Gould on activities in CORE and SDS, Box 122, Folders 10–12, and material from Cuba sit-in, Box 123, Folders 11–13, both in Gould Papers.

3. Material, including photographs, from Gegner barbershop sit-in, Box 122, Folder 8, Gould Papers.

4. Stephen Jay Gould to Joel and Dick, 15 March 1964, Box 122, Folder 6, Gould Papers.

5. Howard Thurman, *Jesus and the Disinherited* (Nashville: Abingdon, 1949), 13–15; James Baldwin, "East River, Downton: Postscript to a Letter from Harlem," in *Nobody Knows My Name* (New York: Dell, 1961), 67–74.

6. Elaine Tyler May, *America and the Pill: A History of Promise, Peril, and Liberation* (New York: Basic Books, 2010); Alondra Nelson, *Body and Soul: The Black Panther Party and the Fight against Medical Discrimination* (Minneapolis: University of Minnesota Press, 2011).

7. Robert Ardrey, *African Genesis: A Personal Investigation into the Animal Origins and Nature of Man* (New York: Atheneum, 1966); Konrad Lorenz, *On Aggression*, trans. Marjorie Kerr (New York: Harcourt, Brace &World, 1966); Desmond Morris, *The Naked Ape: A Zoologist's Study of the Human Animal* (New York: McGraw-Hill, 1967).

8. Erika Lorraine Milam, *Creatures of Cain: The Hunt for Human Nature in Cold War America* (Princeton, NJ: Princeton University Press, 2019), 2.

9. Audra J. Wolfe, *Freedom's Laboratory: The Cold War Struggle for the Soul of Science* (Baltimore: Johns Hopkins University Press, 2018), 174–76.

10. Aaron Panofsky, *Misbehaving Science: Controversy and the Development of Behavior Genetics* (Chicago: University of Chicago Press, 2014), 76.

11. *Racial Isolation in the Public Schools: A Report of the United States Commission on Civil Rights* (1967), 1:138, quoted in Arthur R. Jensen, "How Much Can We Boost IQ and Scholastic Achievement?," *Harvard Educational Review* 39, no. 1 (1969): 3. Alongside the federally funded Project Head Start, Jensen also cited the Banneker Project in St. Louis and Higher Horizons in New York as "failures" of compensatory education programs.

12. Richard Kluger, *Simple Justice: The History of* Brown v. Board of Education *and Black America's Struggle for Equality*, First Vintage Books Edition (New York: Random House, 2004), 354.

13. Jensen, "How Much Can We Boost IQ and Scholastic Achievement?," 79.

14. Jensen, 82.

15. Bill Zimmerman, Len Radinsky, Mel Rothenberg, and Bart Meyers, "Toward a Science for the People," in *Science for the People: Documents from America's Movement of Radical Scientists*, ed. Sigrid Schmalzer, Daniel S. Chard, and Alyssa Botelho (Amherst: University of Massachusetts Press, 2018), 21.

16. Zimmerman et al., 22.

17. Zimmerman et al., 23.

18. Kelly Moore, *Disrupting Science: Social Movements, American Scientists, and the Politics of the Military, 1945–1975* (Princeton, NJ: Princeton University Press, 2009), 161–62.

19. Richard Lewontin and Richard Levins, *Biology Under the Influence: Dialectical Essays on Ecology, Agriculture, and Health* (New York: Monthly Review Press, 2007), 10.

20. Richard C. Lewontin, "Race and Intelligence," *Bulletin of the Atomic Scientists* 26, no. 3 (1970): 6; Lewontin, "Further Remarks on Race and the Genetics of Intelligence," *Bulletin of the Atomic Scientists* 26, no. 5 (1970): 23–24.

21. Lewontin, "Further Remarks on Race and the Genetics of Intelligence," 25.

22. Lewontin, 24.

23. Lewontin, 7.

24. Lewontin, 23.

25. Lewontin, 24.

26. Lewontin, 8.

27. Edward O. Wilson, "The Attempt to Suppress Human Behavioral Genetics," *Journal of General Education* 29, no. 4 (1978): 280–81.

28. Arthur R. Jensen, "Race and the Genetics of Intelligence: A Reply to Lewontin," *Bulletin of the Atomic Scientists* 26, no. 5 (1970): 17.

29. Jensen, 17.

30. Jensen, 22–23.
31. Jensen, 17.
32. Moore, *Disrupting Science*, 2.
33. Gould wrote an account of his family history for his two sons. He knew that both sets of his grandparents were Eastern European Jewish immigrants to New York City, but he knew a great deal more about his maternal grandmother's immigration from Hungary through Austria to work as a seamstress in a "sweatshop in the lower east side." Stephen Jay Gould, "For Jesse and Ethan from Daddy," 26 December 1977, Box 667, Folder 8, Gould Papers.
34. Leonard Gould to Stephen Jay Gould, [after 1965], Box 157, Folder 3, Gould Papers.
35. Stephen Jay Gould, "In Praise of Charles Darwin," *Discover*, February 1982, 20.
36. Jenny Johnson, "Guide to the Stephen Jay Gould Papers," August 2011, revised 2019, Department of Special Collections, Stanford University Libraries, Stanford, CA.
37. Leonard Gould death certificate, 19 October 1981, Box 663, Folder 3, Gould Papers; funeral eulogy for Leonard Gould, ca. 1981 (from death certificate), Box 663, Folder 3, Gould Papers.
38. Leonard Gould, "FBI File," 12 February 1968, Box 663, Folder 6, Gould Papers.
39. Leonard Gould to Eugene Sharpe, 28 October 1965, Box 663, Folder 8, Gould Papers.
40. Leonard Gould to Ashley Montagu, 25 November 1964, Box 664, Folder 7, Gould Papers.
41. Letters by Leonard Gould, 1964–67, Box 664, Folder 7, Gould Papers.
42. Stephen Jay Gould, *The Structure of Evolutionary Theory* (Cambridge, MA: Harvard University Press, 2002), 1018.
43. James Miller and Jim Miller, *Democracy Is in the Streets: From Port Huron to the Siege of Chicago* (Cambridge, MA: Harvard University Press, 1987); Kevin Mattson, *Intellectuals in Action: The Origins of the New Left and Radical Liberalism, 1945–1970* (University Park: Pennsylvania State University Press, 2002); Doug Rossinow, "'The Revolution Is about Our Lives': The New Left's Counterculture," in *Imagine Nation: The American Counterculture in the 1960s and '70s*, ed. Peter Braunstein and Michael William Doyle (New York: Routledge, 2002), 99–124; John Campbell McMillian and Paul Buhle, *The New Left Revisited* (Philadelphia: Temple University Press, 2003); Van Gosse, *Rethinking the New Left: An Interpretative History* (New York: Palgrave Macmillan, 2005).
44. Literature kept by Gould on activities in CORE and SDS.
45. Material from Cuba social activism, Box 123, Folders 12–13, Gould Papers.
46. Literature kept by Gould on activities in CORE and SDS.
47. James Farmer, interview by Robert Penn Warren, 11 June 1964, Robert Penn Warren's Who Speaks for the Negro?, Tape #1, Transcript, 1–2, https://whospeaks.library.vanderbilt.edu/interview/james-farmer-jr; Joan Singler, Jean C. Durning, Bettylou Valentine, and Martha (Maid) J. Adams, *Seattle in Black and White: The Congress of Racial Equality and the Fight for Equal Opportunity* (Seattle: University of Washington Press, 2011), vii.
48. August Meier and Elliott Rudwick, *CORE: A Study in the Civil Rights Movement, 1942–1968* (New York: Oxford University Press, 1973), 39, 159–210.

49. “Memorial March and Gegner Barber Shop Controversy in Yellow Springs Ohio,” 30 September 1963, WYSO 91.3 FM Public Radio Digital Archives, Greene County Room Digital Collections, Greene County Public Library, https://cdm17207.contentdm.oclc.org/digital/collection/WYSOProgram/id/196.

50. Farmer interview, Tape #1, Transcript, 2–3.

51. W. E. B. Du Bois, *The Talented Tenth* (New York: James Pott, 1903).

52. Stephen Jay Gould, “Pleistocene and Recent History of the Subgenus Poecilozonites (*Poecilozonites*) (Gastropoda: Pulmonata) in Bermuda: An Evolutionary Microcosm” (PhD diss., Columbia University, 1967).

53. Audra J. Wolfe, *Competing with the Soviets: Science, Technology, and the State in Cold War America* (Baltimore: Johns Hopkins University Press, 2013), 105–15.

54. John T. Bethell, *Harvard Observed: An Illustrated History of the University in the Twentieth Century* (Cambridge, MA: Harvard University Press, 1998), 220.

55. Quoted in Bethell, 220.

56. Bethell, 221.

57. Bethell, 232 and 225.

58. George Gaylord Simpson, *Tempo and Mode in Evolution* (New York: Columbia University Press, 1944).

59. Stephen Jay Gould to Alan Ternes, 25 June 1973, Box 230, Folder 3, Gould Papers.

60. “Memorial March and Gegner Barber Shop Controversy”; “Ohio Barber Shop Shuts in Protests.”

61. Nishani Frazier, *Harambee City: The Congress of Racial Equality in Cleveland and the Rise of Black Power Populism* (Fayetteville: University of Arkansas Press, 2017), xvii–xviii.

62. Gould wrote about his dislike for the work of organizing to Tom Schopf regarding their collaboration in paleobiology in a letter of October 1973: “I just state my preferences in characteristic prose and shun the dirty work of organization that seems so difficult for me.” Stephen Jay Gould to Tom Schopf, 5 October 1973, Box 5, Folder 14, Thomas J. M. Schopf Papers, Smithsonian Institution Archives, Record Unit 7429. In this letter he refers to the work of organizing a group of scientists, but the sentiment was broadly echoed throughout his correspondence. Moreover, Gould tended to disavow his association with specific New Left organizations, preferring to be considered as an individual. When asked during his deposition for the McLean trial whether he was a member of Science for the People, he replied, “Yes—I am a member insofar as I pay my dues to get the magazine. That’s my only contact with it.” “Deposition of Stephen Jay Gould, 27 November 1981,” McLean v. Arkansas Documentation Project, 139, accessed 10 June 2013, https://www.antievolution.org/projects/mclean/new_site/depos/pf_gould_dep.htm.

63. Martin Luther King Jr., *Letters to a Birmingham Jail: A Response to the Words and Dreams of Dr. Martin Luther King, Jr.*, ed. Bryan Loritts, new ed. (Chicago: Moody, 2014), 37.

64. King, 37. Writing in 1991, the civil rights historian Steven Lawson argued that the country’s then period of conservative retrenchment fueled a desire to memorialize the era of Black protest and activism in the palatable terms of racial universalism. This desire was felt particularly in the remembrances of Martin Luther King Jr., who Lawson

claimed emerged as "a perennial dreamer, frozen in time at his most famous address during the 1963 March on Washington." Steven F. Lawson, "Freedom Then, Freedom Now: The Historiography of the Civil Rights Movement," *American Historical Review* 96, no. 2 (1991): 460.

65. Gould, "Racist Arguments and I.Q.," 24.

66. Gould, 29.

67. Stephen Jay Gould, "The Race Problem," *Natural History* 83, no. 3 (1974): 8.

68. Stephen Jay Gould, *Ever Since Darwin: Reflections in Natural History* (New York: Norton, 1977), 231.

69. Richard C. Lewontin, "The Apportionment of Human Diversity," in *Evolutionary Biology: Volume 6*, ed. Theodosius Dobzhansky, Max K. Hecht, and William C. Steere (New York: Springer US, 1972), 397.

70. Beverly Smith, interview, in *How We Get Free: Black Feminism and the Combahee River Collective*, ed. Keeanga-Yamahtta Taylor (Chicago: Haymarket Books, 2017), 71–109.

71. Combahee River Collective, "The Combahee River Collective Statement, 1977," in Keeanga-Yamahtta Taylor, *How We Get Free*, 18.

72. Smith interview, 97.

73. Combahee River Collective, "Combahee River Collective Statement, 1977," 26.

74. Combahee River Collective, 21.

75. Nelson, *Body and Soul*, xii.

76. Moore, *Disrupting Science*, 181.

77. Kelly Moore argues that by the mid-1970s feminists had begun to have a more prominent role in SftP, as evidenced by their greater visibility in the magazine. Moore, 182–85.

78. Derrick A. Bell, "Brown v. Board of Education and the Interest-Convergence Dilemma," *Harvard Law Review* 93, no. 3 (1980): 523.

79. Bell, 524.

80. Cornel West, *Prophesy Deliverance! An Afro-American Revolutionary Christianity* (Philadelphia: Westminster, 1982), 47.

81. Richard Nixon, "Republican National Convention Speech, 1972," quoted in *A War for the Soul of America: A History of the Culture Wars*, by Andrew Hartman (Chicago: University of Chicago Press, 2015), 106.

82. Moore, *Disrupting Science*, 166.

83. See, e.g., the list of members of the advisory board in *Science for the People* 17, no. 4 (1985).

84. Gould discussed his approach to humanistic natural history writing as "Galilean," delighting in the "intellectual puzzles of nature," in Stephen Jay Gould, *Bully for Brontosaurus: Reflections in Natural History* (New York: Norton, 1991), 11–13.

85. Stephen Jay Gould, *The Panda's Thumb: More Reflections in Natural History* (New York: Norton, 1980), 13.

86. Stephen Jay Gould, "The Child as Man's Real Father," *Natural History* 84, no. 5 (May 1975): 18.

87. Gould, 22.

88. Gillian Fuller, "Cultivating Science: Negotiating Discourse in the Popular Texts

of Stephen Jay Gould," in *Reading Science: Critical and Functional Perspectives on Discourses of Science, ed.* J. R. Martin and Robert Veel (New York: Psychology Press, 1998), 35–38.

89. "This View of Stephen Jay Gould," *Natural History* 108, no. 9 (2002): 48–49.

90. W. Drake McFeeley, *Books That Live: Norton's First One Hundred Years* (New York: Norton, 2023).

91. Gould, *Ever Since Darwin*; Gould, *Panda's Thumb*; Stephen Jay Gould, *The Mismeasure of Man* (New York: Norton, 1981); Gould, *Hen's Teeth and Horse's Toes: Further Reflections in Natural History* (New York: Norton, 1983); Gould, *The Flamingo's Smile: Reflections in Natural History* (New York: Norton, 1985); Rosamund W. Purcell and Stephen Jay Gould, *Illuminations: A Bestiary* (New York: Norton, 1986); Gould, *Wonderful Life: The Burgess Shale and the Nature of History* (New York: Norton, 1989); Gould, *Bully for Brontosaurus*; Rosamond Wolff Purcell and Stephen Jay Gould, *Finders, Keepers: Eight Collectors* (New York: Norton, 1992); Stephen Jay Gould, *Eight Little Piggies: Reflections in Natural History* (New York: Norton, 1993).

92. Norton publicity clipping for *Ever Since Darwin*, Box 150, Folder 5, Gould Papers. Gould's relationship with Norton ended after the publication of *Eight Little Piggies in* 1993. It was the sixth anthology of "This View of Life."

93. Harold Hayes, "Seeking Our Roots: A Few Million Years Ago," *Chicago Sun Times*, 15 January 1978, Box 150, Folder 4, Gould Papers.

94. Norman Macbeth, "Review of *Ever Since Darwin: Reflections on Natural History* by Stephen Jay Gould," *Systematic Zoology* 27, no. 2 (1978): 264–66.

95. Edwin Barber to Stephen Jay Gould, 25 April 1978, Box 150, Folder 4, Gould Papers.

96. Some examples of reviews are Raymond A. Sokolov, "Talk With Stephen Jay Gould," *New York Times*, 20 November 1977, sec. BR, 4; Frank Lipsius, "The Incontestable Fascination of Real Nature," *Philadelphia Inquirer*, 5 March 1978; and "Notes on Current Books," *Virginia Quarterly 54, no. 2 (*1978): 70. See also "Book Reviews of *Ever Since Darwin*," Box 150, Folder 4, Gould Papers.

97. Kurt Vonnegut to Stephen Jay Gould, 10 October 1985, Box 698, Folder 3, Gould Papers.

98. Stephen Jay Gould to Peter Stone, 23 November 1979, Box 120, Folder 18, Gould Papers.

99. Gould, *Ever Since Darwin*, 16.

100. Michelle Alexander, *The New Jim Crow: Mass Incarceration in the Age of Colorblindness* (New York: New Press, 2010), 8.

101. Alexander, 6.

102. At the time of this writing, forty-eight states and the District of Columbia prevent people from voting while incarcerated, and in most states this continues after parole, in some cases even after the end of the sentence, for a period of one year up to the rest of their life. See Alexander, 196–97.

103. Michelle Alexander, "Injustice on Repeat," op-ed, *New York Times, Late Edition (East Coast)*, 19 January 2020, sec. SR.

104. Keeanga-Yamahtta Taylor, *From #BlackLivesMatter to Black Liberation* (Chicago: Haymarket Books, 2016), 2–12.

105. Alexander, "Injustice on Repeat."

106. Keeanga-Yamahtta Taylor, "The Pivot to Class," in *Fifty Years Since MLK*, ed. Deborah Chasman and Joshua Cohen (Cambridge, MA: Boston Review, 2017), 36–39.

107. Taylor, 37.

108. "'A Testament of Hope' by Dr. King," Hef's Philosophy: Playboy and Revolution from 1965–1975, accessed 28 July 2022, https://forthearticles.omeka.net/items/show/37, quoted in Taylor, "Pivot to Class," 36.

109. Frazier, *Harambee City*, 30–32, 109.

110. City News Service, "Study: Deaths of Black and Latino Men in LA County Jails Misclassified," *Spectrum News 1*, 1 June 2022, https://spectrumnews1.com/ca/la-west/public-safety/2022/06/01/study--deaths-of-black-and-latino-men-in-la-county-jails-misclassified#:~:text=LOS%20ANGELES%20(CNS)%20%E2%80%94%20The,guards%2C%20UCLA%20researchers%20claimed%20Wednesday.

111. "Coroner Reports Project: Death Under the U.S. Medical Examiner / Coroner System," Terence Keel, accessed 28 July 2022, https://www.terencekeel.com/coroner-reports-projects; "About Us," Dignity and Power Now, accessed 28 July 2022, https://dignityandpowernow.org/about-us/.

112. Ruha Benjamin, *Race After Technology: Abolitionist Tools for the New Jim Code* (Cambridge: Polity Press, 2019), 8.

113. Ruha Benjamin, "Black AfterLives Matter: Cultivating Kinfulness as Reproductive Justice," *Boston Review*, 16 July 2018, https://www.bostonreview.net/articles/ruha-benjamin-black-afterlives-matter/.

Chapter 2 • Sociobiology and Gender Trouble

Epigraph: Stephen Jay Gould, "The Nonscience of Human Nature," *Natural History Magazine* 83, no. 4 (May 1974): 24.

1. Stephen Jay Gould, "Biological Potential vs. Biological Determinism," *Natural History* 85, no. 5 (1976): 12–22; Edward O. Wilson, *Sociobiology: The New Synthesis* (Cambridge, MA: Belknap Press of Harvard University Press, 1975).

2. Wilson, *Sociobiology*, 547–74.

3. Erika Lorraine Milam, *Creatures of Cain: The Hunt for Human Nature in Cold War America* (Princeton, NJ: Princeton University Press, 2019), 239–42. For the works by Ardrey, Lorenz, and Morris, see chap. 1, n. 7.

4. Richard Dawkins, *The Selfish Gene* (New York: Oxford University Press, 1976).

5. Randy Thornhill and Craig T. Palmer, *A Natural History of Rape: Biological Bases of Sexual Coercion* (Cambridge, MA: MIT Press, 2000); Steven Pinker, *The Blank Slate: The Modern Denial of Human Nature* (New York: Penguin Books, 2003), 337–71.

6. Meeting minutes from Sociobiology Study Group, April 1977, in Sociobiology Study Group minutes, 1975–78, folder 119, Papers of Freda Friedman Salzman, 1927–1981, Schlesinger Library, Radcliffe Institute, Harvard University, Cambridge, MA, https://hollisarchives.lib.harvard.edu/repositories/8/resources/9171.

7. Jonathan R. Beckwith, *Making Genes, Making Waves: A Social Activist in Science* (Cambridge, MA: Harvard University Press, 2009), 5–7.

8. "Biographical Note, Collection Finding Aide," n.d., Papers of Freda Friedman Salzman, 1927–1981.

9. According to the meeting minutes of the group, Gould began attending SSG

meetings on 22 June 1976. Minutes of SSG meeting, 22 June 1976, Sociobiology Study Group minutes, 1975–1978, folder 119, Papers of Freda Friedman Salzman, 1927–1981.

10. Nina McCain, "Sociobiology: New Theory on Man's Motivation," *Boston Globe,* 13 July 1975, A1; John Pfeiffer, "Sociobiology," *New York Times Book Review,* 27 July 1975, sec. BR, 4.

11. C. H. Waddington, "Mindless Societies," *New York Review of Books,* 7 August 1975. The SSG's critical response to Waddington's review has left the impression that Waddington was in support of Wilson's *Sociobiology.* Elizabeth Allen, Barbara Beckwith, Jon Beckwith, Steven Chorover, David Culver, Margaret Duncan, Steven Goud, Ruth Hubbard, Hiroshi Inouye, Anthony Leeds, Richard Lewontin, Chuck Madansky, Larry Miller, Reed Pyertiz, Peter Bent Brigham, Miriam Rosenthal, and Herb Schreier, "Against Sociobiology," *New York Review of Books,* 13 November 1975. However, Erik L. Peterson has argued that when Waddington's review is read in the fuller context of his career (and the essay itself), one sees that Waddington was in fact critical of Wilson's "gene-centric mechanism." Peterson, *The Life Organic: The Theoretical Biology Club and the Roots of Epigenetics* (Pittsburgh: University of Pittsburgh Press, 2016), 230. Further, Waddington passed away on 26 September 1975, so he was unable to reply to the letter from Allen et al.

12. Allen et al., "Against Sociobiology."

13. John Beckwith, interview by author, 21 July 2016, Boston; Beckwith, *Making Genes, Making Waves,* 142.

14. For a discussion of these protest activities as documented in the archive, see Neil Jumonville, "The Cultural Politics of the Sociobiology Debate," *Journal of the History of Biology* 35, no. 3 (2002): 582.

15. Wilson, *Sociobiology,* 553.

16. In nineteenth-century accounts of evolution, civilization was released from the primitive past. Certainly, some categories in society (i.e., women, nonwhite races, and mental degenerates) were depicted as being arrested in their evolutionary development, but civilization itself was a quality that had moved beyond the earlier, primitive state. For further discussion, see, e.g., Gail Bederman, *Manliness and Civilization: A Cultural History of Gender and Race in the United States, 1880–1917* (Chicago: University of Chicago Press, 1995); and Laura L. Lovett, *Conceiving the Future: Pronatalism, Reproduction, and the Family in the United States, 1890–1938* (Chapel Hill: University of North Carolina Press, 2007).

17. McCain, "Sociobiology"; Robert M. May, "Sociobiology: A New Synthesis and an Old Quarrel," *Nature* 260, no. 5550 (1976): 390–92; Nicholas Wade, "Sociobiology: Troubled Birth for New Discipline," *Science* 191, no. 4232 (1976): 1151; Myra MacPherson, "Sociobiology: Scientists at Odds; How Much of Our Behavior Is Determined by Our Genes?," *Washington Post,* 21 November 1976, B1; Joel Velasco, "Why You Do What You Do: Sociobiology; A New Theory of Behavior," *Time,* 1 August 1977.

18. Both Lewontin and Gould stopped attending meetings after May 1977, as noted in Jumonville, "Cultural Politics of the Sociobiology Debate," 573. Even after that, though, members of the group believed that Gould's efforts were in keeping with their cause. Beckwith interview. Gould continued to receive the meeting minutes of the SSG through the 1980s. Sociobiology Study Group Folders, Box 889, Folder 4, Gould Papers.

19. Myra MacPherson to Bob [after 21 November 1976], Sociobiology Study Group minutes, 1975–1978, folder 119, Papers of Freda Friedman Salzman, 1927–1981.

20. This frustration is apparent in the minutes from the early meetings of the group. Sociobiology Study Group minutes, 1975–1978, folder 118, Papers of Freda Friedman Salzman, 1927–1981.

21. Edward O. Wilson, *Naturalist* (Washington, DC: Island Press, 2006), 339.

22. Minutes of SSG meeting, 10 May 1977, Sociobiology Study Group minutes, 1975–1980, Papers of Freda Friedman Salzman.

23. Jumonville, "Cultural Politics of the Sociobiology Debate," 569–70.

24. Wilson, *Sociobiology*, 547.

25. Dawkins, *Selfish Gene*, 151–78.

26. Kimberly A. Hamlin, *From Eve to Evolution: Darwin, Science, and Women's Rights in Gilded Age America* (Chicago: University of Chicago Press, 2014), 57–93.

27. Sandra Morgen, *Into Our Own Hands: The Women's Health Movement in the United States, 1969–1990* (New Brunswick, NJ: Rutgers University Press, 2002), 3–15.

28. Nadine Weidman, *Killer Instinct: The Popular Science of Human Nature in Twentieth-Century America* (Cambridge, MA: Harvard University Press, 2021), 223–59.

29. "Genes and Gender: A Symposium," 29 January 1977, MC 526, Box 1, Folder 1, Records of the Genes and Gender Collective, 1974–1999, Schlesinger Library, Radcliffe Institute, Harvard University, Cambridge, MA.

30. "Participant Responses, Genes and Gender Conference I," 1977, MC 526, Box 1, Folder 1, Records of the Genes and Gender Collective, 1974–1999.

31. Ethel Tobach and Betty Rosoff, eds., *Genes and Gender I* (New York: Gordian, 1978), 11–12.

32. Anne Briscoe, "Hormones and Gender," in Tobach and Rosoff, *Genes and Gender I*, 45.

33. "AWIS History," Association for Women in Science, accessed 26 July 2022, https://www.awis.org/about-awis/awis-history/.

34. Tiffany K. Wayne, *American Women of Science since 1900* (Santa Barbara, CA: ABC-CLIO, 2011), 252–53.

35. Christine Ward Gailey, "Eleanor Burke Leacock and Historical Transformations of Gender: Beyond Timeless Patriarchy," in "Genealogies of the Feminist Present: Lineages and Connections in Feminist Anthropology," ed. Lynn Bolles and Mary H. Moran, American Ethnologist Website, 24 May 2021, https://americanethnologist.org/features/collections/legacies-and-genealogies-in-feminist-anthropology/eleanor-burke-leacock-and-historical-transformations-of-gender-beyond-timeless-patriarchy.

36. Eleanor Burke Leacock, *The Culture of Poverty: A Critique* (New York: Simon & Schuster, 1971).

37. Eleanor Leacock, "Society and Gender," in Tobach and Rosoff, *Genes and Gender I*, 79–80. Leacock's chapter focused particularly on E. O. Wilson, "Human Decency Is Animal," *New York Times Magazine*, 12 October 1975, 272.

38. Leacock, "Society and Gender," 79.

39. Tobach and Rosoff, *Genes and Gender I*; Jenna Tonn, "Radical Science, Feminism, and the Biology of Determinism," *Lady Science* (blog), 20 December 2018, https://thenewinquiry.com/blog/radical-science-feminism-and-the-biology-of-determinism/.

40. Marian Lowe and Ruth Hubbard to Ethel Tobach and Betty Rosoff, 2 September 1979, MC 526, Box 2, Folder 5, Records of the Genes and Gender Collective, 1974–1999.

41. Beryl Lieff Benderly, "Ruth Hubbard and the Evolution of Biology," Science, Taken For Granted, 5 October 2016, http://www.science.org/content/article/ruth-hubbard-and-evolution-biology.

42. Ruth Hubbard, *The Politics of Women's Biology* (New Brunswick, NJ: Rutgers University Press, 1990), 1.

43. Ruth Hubbard and Marian Lowe, eds., *Genes and Gender II: Pitfalls in Research on Sex and Gender* (New York: Gordian, 1979), 13.

44. Gould, "Nonscience of Human Nature," 21.

45. Hubbard and Lowe, *Genes and Gender II*, 19.

46. Hubbard and Lowe, 15n5.

47. Benderly, "Ruth Hubbard and the Evolution of Biology."

48. Ruth Hubbard, "Have Only Men Evolved?," in *Women Look at Biology Looking at Women: A Collection of Feminist Critiques, ed. Ruth Hubbard, Mary Sue Henifin, and Barbara Fried* (Boston: G. K. Hall, 1979), 7–36.

49. Hubbard, 32.

50. Ruth Hubbard, with Wendy Sandford, "New Reproductive Technologies," in Boston Women's Health Book Collective, *The New Our Bodies, Ourselves: A Book by and for Women* (New York: Simon & Schuster, 1984), 317–24.

51. Boston Women's Health Book Collective, *Our Bodies, Ourselves: A Book by and for Women* (Boston, 1970).

52. Kathy Davis, *The Making of "Our Bodies, Ourselves": How Feminism Travels across Borders* (Durham, NC: Duke University Press, 2007), 1–3.

53. Jenna Tonn, "Fight over the Technological Future of Motherhood," 28 August 2020, The Immanent Frame: Secularism, Religion, and the Public Sphere, https://tif.ssrc.org/2020/08/28/fight-over-the-technological-future-of-motherhood/.

54. See, e.g., E. O. Wilson to Stephen Jay Gould, 4 March and 24 May 1974, Box 230, Folder 2, Gould Papers.

55. Gould, "Biological Potential vs. Biological Determinism," 12.

56. Gould, 22.

57. Gould, 18.

58. Gould, 18–19.

59. Gould, 19.

60. Gould, 20.

61. Gould to Wilson, 16 March 1976, Box 230, Folder 2, Gould Papers.

62. Wilson to Gould, 16 March 1976. The articles Wilson referred to in his letter to Gould were May, "Sociobiology"; Wade, "Sociobiology"; and Sociobiology Study Group of Science for the People, "Dialogue. The Critique: Sociobiology; Another Biological Determinism," *BioScience* 26, no. 3 (1976): 182–86. Wilson responded to the last review in Edward O. Wilson, "Dialogue. The Response: Academic Vigilantism and the Political Significance of Sociobiology," *BioScience* 26, no. 3 (1976): 183–90.

63. Wilson to Gould, 16 March 1976.

64. Stephen Jay Gould to E. O. Wilson, 17 March 1976, Box 230, Folder 2, Gould Papers.

65. Draft of Wilson's reply sent to Alan Ternes (attached to letter to Alan Ternes dated 5 April 1976) and copied to Stephen Jay Gould, 1976, Box 230, Folder 2, Gould Papers.

66. Manuscript draft of Stephen Jay Gould's reply to Edward O. Wilson, [1976], Box 230, Folder 2, Gould Papers.

67. Manuscript draft of Gould's reply to Wilson, [1976].

68. The Committee Against Racism, or CAR, was an organization of the Progressive Labor Party founded in 1973. Harvey Klehr, *Far Left of Center: The American Radical Left Today* (New Brunswick, NJ: Transaction, 1991), 189–91. For an account of the water dousing see, Weidman, *Killer Instinct*, 224.

69. Edward O. Wilson to Stephen Jay Gould, 27 February 1978, Box 230, Folder 2, Gould Papers.

70. Garland Allen and Stephen Jay Gould, "Book proposal for Biological Determinism for Norton Inc.," Box 149, Folder 8, Gould Papers.

71. Garland Allen to Stephen Jay Gould, 23 February 1977, Box 1, Folder 4, Gould Papers.

72. Allen wrote to Gould on 16 February 1979 that "one of my friends of Ann Arbor SftP expressed regret that our project folded, and wanted to see if we could patch things up between us. I said I was amenable, but doubted that our views could be more easily reconciled by 'arbitrators' than by us ourselves." Box 1, Folder 4, Gould Papers.

73. Garland E. Allen, *Thomas Hunt Morgan: The Man and His Science* (Princeton, NJ: Princeton University Press, 1978); Allen, *Life Science in the Twentieth Century*, Cambridge History of Science (London: Cambridge University Press, 1979); Allen, "The Misuse of Biological Hierarchies: The American Eugenics Movement, 1900–1940," *History and Philosophy of the Life Sciences* 5, no. 2 (1983): 105–28.

74. Stephen Jay Gould. *The Mismeasure of Man*. (New York: Norton, 1981), 28.

75. Gould, 326.

76. Gould, 330–31.

77. "Norton Inc. Book Release: The Mismeasure of Man," Box 354, Folder 2, Gould Papers.

78. Gould, *Mismeasure of Man*, 21.

79. Brian Regal, *Henry Fairfield Osborn: Race, and the Search for the Origins of Man* (Farnham, UK: Ashgate, 2002), xii–xix.

80. Stephen Jay Gould, *Ontogeny and Phylogeny* (Cambridge, MA: Harvard University Press, 1977), 7.

81. Gould, 123.

82. Gould, *Mismeasure of Man*, 124.

83. Gould's work in *Ontogeny and Phylogeny* is cited as an important inspirational early moment in the rise of the modern field of evolutionary developmental biology, or evo-devo. See, e.g., Brian K. Hall, "Evo-Devo: Evolutionary Developmental Mechanisms," *International Journal of Developmental Biology* 47, no. 7/8 (2003): 491; Alan C. Love, "Evolutionary Morphology, Innovation, and the Synthesis of Evolutionary and Developmental Biology," *Biology & Philosophy* 18, no. 2 (2003): 310; and Roger Jankowski, *The Evo-Devo Origin of the Nose, Anterior Skull Base and Midface* (New York: Springer Science & Business Media, 2013), 1.

84. Gould, *Ontogeny and Phylogeny*, 6.

85. See Thomas S. Kuhn, *The Structure of Scientific Revolutions* (Chicago: University of Chicago Press, 1962). For a prominent example of Gould's reflection on Kuhnian paradigms, see Stephen Jay Gould, *The Structure of Evolutionary Theory* (Cambridge, MA: Harvard University Press, 2002), 960–72. For a holistic discussion of Kuhn's influence on Gould, see Gregory Radick, "The Exemplary Kuhnian: Gould's Structure Revisited," *Historical Studies in the Natural Sciences* 42, no. 2 (2012): 143–57.

86. Samuel George Morton, *Crania Americana; or, A comparative view of the skulls of various aboriginal nations of North and South America* (Philadelphia: J. Dobson, 1839). Ann Fabian explains that while Morton was reluctant to definitively commit himself to the polygenist position, that position was mirrored in his arguments that the different races represented different species of humanity. Ann Fabian, *The Skull Collectors: Race, Science, and America's Unburied Dead* (Chicago: University of Chicago Press, 2010), 83.

87. "*The Mismeasure of Man*, research and notes," Box 264, Folder 3, Gould Papers.

88. Gould, *Mismeasure of Man*, 54–56.

89. Cynthia Eagle Russett, *Sexual Science: The Victorian Construction of Womanhood* (Cambridge, MA: Harvard University Press, 1989), 33nn31 and 32, 60n34.

90. Margalit Fox, "Cynthia Eagle Russett, Chronicler of Women's History, Dies at 76," *New York Times*, 18 December 2013, Books, https://www.nytimes.com/2013/12/19/books/cynthia-russett-historian-of-women-dies-at-76.html; Janet Horowitz Murray, "When Sexism Was Science," review of Russett, *Sexual Science*, *New York Times*, 9 April 1989.

91. Londa Schiebinger, *Nature's Body: Gender in the Making of Modern Science* (Boston: Beacon, 1993), 79, 235n12.

92. Marla R. Miller, "Tracking the Women's Movement through the Women's Action Alliance," *Journal of Women's History* 14, no. 2 (2002): 154–56.

93. "Finding Aid for Women's Action Alliance Records, Series II. Projects, 1972–96," Smith College, Libraries, Sophia Smith Collection of Women's History, accessed 26 July 2022, https://findingaids.smith.edu/repositories/2/archival_objects/151371.

94. Russett, *Sexual Science*, 14.

95. Russett, 15.

96. Schiebinger, *Nature's Body*, 10.

97. Sarah S. Richardson, "Feminist Philosophy of Science: History, Contributions, and Challenges," *Synthese: An International Journal for Epistemology, Methodology and Philosophy of Science* 177, no. 3 (2010): 339.

98. Catharine R. Stimpson and Joan N. Burstyn, editorial, in "Women, Science, and Society," ed. Stimpson and Burstyn, special issue, *Signs: Journal of Women in Culture and Society* 4, no. 1 (1978): 1.

99. See Sally Gregory Kohlstedt, *The Formation of the American Scientific Community: The American Association for the Advancement of Science, 1848–60* (Champaign: University of Illinois Press, 1976); Kohlstedt, "In from the Periphery: American Women in Science, 1830–1880," in Stimpson and Burstyn, "Women, Science, and Society," special issue, *Signs* 4, no. 1 (1978): 81–96; Kohlstedt, "Parlors, Primers, and Public Schooling: Education for Science in Nineteenth-Century America," *Isis* 81, no. 3 (1990): 424–45; Donna J. Haraway, *Primate Visions: Gender, Race, and Nature in the World of Modern Science* (New York: Routledge, 1989); and Haraway, *Simians, Cyborgs, and Women: The Reinvention of Nature* (New York: Routledge, 1991). On the influences of

these figures, see https://hssonline.org/page/sartonmedal, on Kohlstedt being awarded the Sarton Medal by the History of Science Society in 2018; and Moira Weigel, "Feminist Cyborg Scholar Donna Haraway: 'The Disorder of Our Era Isn't Necessary,'" *Guardian*, 20 June 2019, World News, https://www.theguardian.com/world/2019/jun/20/donna-haraway-interview-cyborg-manifesto-post-truth.

100. Marian Lowe, "Sociobiology and Sex Differences," in Stimpson and Burstyn, "Women, Science, and Society," special issue, *Signs* 4, no. 1 (1978): 119.

101. Kohlstedt, "In from the Periphery," 96.

102. Donna Haraway, "Animal Sociology and a Natural Economy of the Body Politic, Part I: A Political Physiology of Dominance," in Stimpson and Burstyn, "Women, Science, and Society," special issue, *Signs* 4, no. 1 (1978): 22–23.

103. Nancy Tuana, introduction to "Feminism and Science," ed. Tuana, special issue, *Hypatia* 2, no. 3 (1987): 1.

104. "Helen Longino," Feminist Theory Website, accessed 1 December 2016, https://www.cddc.vt.edu/feminism/Longino.html.

105. Helen E. Longino, "Can There Be A Feminist Science?," in Tuana, "Feminism and Science," special issue, *Hypatia* 2, no. 3 (1987): 52.

106. Longino, 53.

107. Longino, 56.

108. Judith Butler, *Gender Trouble: Feminism and the Subversion of Identity* (New York: Routledge, 1990), xi.

109. Butler, 16.

110. Butler, 151n9.

111. Alexis De Veaux, *Warrior Poet: A Biography of Audre Lorde* (New York: Norton, 2004), 246–53; Lester C. Olson, "The Personal, the Political, and Others: Audre Lorde Denouncing 'The Second Sex Conference,'" *Philosophy & Rhetoric* 33, no. 3 (2000): 259–85.

112. Simone de Beauvoir, *The Second Sex* [1949], trans. H. M. Parshley (New York: Knopf, 1993), 137.

113. Olson, "Personal, the Political, and Others," 259–60.

114. De Veaux, *Warrior Poet*, 239–44.

115. Audre Lorde, "The Master's Tools Will Never Dismantle the Master's House," in *Sister Outsider: Essays and Speeches* (Berkeley, CA: Crossing, 1984), 110–13.

116. Gould, *The Structure of Evolutionary Theory*, 31.

117. Allen et al., "Against Sociobiology."

118. Hubbard and Lowe, *Genes and Gender II*, 19.

119. Stephen Jay Gould, *The Mismeasure of Man*, rev. and expanded ed. (New York: Norton, 1996), 27.

120. See Michel Foucault, *"Society Must Be Defended": Lectures at the College de France, 1975–76*, ed. Mauro Bertani and Alessandro Fontana, trans. David Macey (New York: Picador, 2003), 243–45; Ann Laura Stoler, *Race and the Education of Desire: Foucault's History of Sexuality and the Colonial Order of Things* (Durham, NC: Duke University Press, 1995); Alys Eve Weinbaum, *Wayward Reproductions: Genealogies of Race and Nation in Transatlantic Modern Thought* (Durham, NC: Duke University Press, 2004); and Myrna Perez Sheldon, "Race and Sexuality: A Secular Theory of Race," in

A Cultural History of Race in the Age of Empire and Nation State, ed. Marina B. Mogilner (London: Bloomsbury Academic, 2021), 149–64.

121. Foucault first addressed the concept of biopower in *The History of Sexuality*. See Michel Foucault, *The History of Sexuality, Volume 1: An Introduction*, trans. Robert Hurley (New York: Vintage Books, 1990), 141–43. He most clearly references race as a product of biopower in Foucault, *"Society Must Be Defended,"* 239–64.

122. Jane Carey, "The Racial Imperatives of Sex: Birth Control and Eugenics in Britain, the United States and Australia in the Interwar Years," *Women's History Review* 21, no. 5 (2012): 733–52.

123. Daylanne K. English, *Unnatural Selections: Eugenics in American Modernism and the Harlem Renaissance* (Chapel Hill: University of North Carolina Press, 2004), 35–64.

124. Lovett, *Conceiving the Future*, 109–30.

125. Leslie J. Reagan, *Dangerous Pregnancies: Mothers, Disabilities, and Abortion in Modern America* (Berkeley: University of California Press, 2010).

126. Diane B. Paul, *Controlling Human Heredity, 1865 to the Present* (Amherst, NY: Humanity Books, 1995); Dana Seitler, "Unnatural Selection: Mothers, Eugenic Feminism, and Charlotte Perkins Gilman's Regeneration Narratives," *American Quarterly* 55, no. 1 (2003): 61–88; Alison Bashford and Philippa Levine, *The Oxford Handbook of the History of Eugenics* (Oxford: Oxford University Press, 2010); Staffan Müller-Wille and Hans-Jörg Rheinberger, *A Cultural History of Heredity* (Chicago: University of Chicago Press, 2012), 95–126.

127. Myrna Perez Sheldon, "A Feminist Theology of Abortion," in *Critical Approaches to Science and Religion*, ed. Myrna Perez Sheldon, Ahmed Ragab, and Terence Keel (New York: Columbia University Press, 2023), 61–62.

128. *Clarence Thomas, Kristina Box, Commissioner, Indiana Department of Health, et al. v. Planned Parenthood of Indiana and Kentucky, Inc., et al.*, No. 18–483 (US Supreme Court, 28 May 2019).

129. Sasha Turner, *Contested Bodies: Pregnancy, Childrearing, and Slavery in Jamaica* (Philadelphia: University of Pennsylvania Press, 2017); Deirdre Benia Cooper Owens, *Medical Bondage: Race, Gender, and the Origins of American Gynecology* (Athens: University of Georgia Press, 2017); Myrna Perez Sheldon, "Breeding Mixed-Race Women for Profit and Pleasure," *American Quarterly* 71, no. 3 (2019): 741–65.

130. Frederick E. Hoxie, *A Final Promise: The Campaign to Assimilate the Indians, 1880–1920* (Lincoln: University of Nebraska Press, 2001); David A. Chang, *The Color of the Land: Race, Nation, and the Politics of Landownership in Oklahoma, 1832–1929* (Chapel Hill: University of North Carolina Press, 2010), 1–14; Anne Gregory, "Competency, Allotment, and the Canton Asylum: The Case of a Muscogee Woman," *Disability Studies Quarterly* 41, no. 4 (2021).

131. Bill Ong Hing, *Defining America through Immigration Policy*, Mapping Racisms (Philadelphia: Temple University Press, 2004), 115–84; Julie M. Weise, "Mexican Nationalisms, Southern Racisms: Mexicans and Mexican Americans in the U.S. South, 1908–1939," *American Quarterly* 60, no. 3 (2008): 749–77; Adrián Félix, "New Americans or Diasporic Nationalists? Mexican Migrant Responses to Naturalization and Implications for Political Participation," *American Quarterly* 60, no. 3 (2008): 601–24.

132. Keeanga-Yamahtta Taylor, ed., *How We Get Free: Black Feminism and the Combahee River Collective* (Chicago: Haymarket Books, 2017), 15.

133. Kimberlé Crenshaw, "Demarginalizing the Intersection of Race and Sex: A Black Feminist Critique of Antidiscrimination Doctrine, Feminist Theory and Antiracist Politics," *University of Chicago Legal Forum*, no. 1 (1989): 139–67.

134. Loretta Ross and Rickie Solinger, *Reproductive Justice: An Introduction* (Oakland: University of California Press, 2017); Zakiya Luna, *Reproductive Rights as Human Rights: Women of Color and the Fight for Reproductive Justice* (New York: New York University Press, 2020).

135. Darrel Enck-Wanzer, *The Young Lords: A Reader* (New York: New York University Press, 2010), 192.

136. Alondra Nelson, *Body and Soul: The Black Panther Party and the Fight against Medical Discrimination* (Minneapolis: University of Minnesota Press, 2011), 68–69.

137. Morgen, *Into Our Own Hands*, 5–7.

138. Mark Borrello and David Sepkoski, "Ideology as Biology," *New York Review of Books*, 5 February 2022, https://www.nybooks.com/daily/2022/02/05/ideology-as-biology/.

Chapter 3 • Evolutionarily Engineered Sexism

Epigraph: Stephen Jay Gould, "The Evolutionary Biology of Constraint," *Daedalus*, 1980, 46.

1. Stephen Jay Gould and Richard C. Lewontin, "The Spandrels of San Marco and the Panglossian Paradigm: A Critique of the Adaptationist Programme," *Proceedings of the Royal Society of London. Series B. Biological Sciences* 205, no. 1161 (1979): 581–98; Gerald Borgia, "The Scandals of San Marco," *Quarterly Review of Biology* 69, no. 3 (1994): 373; Massimo Pigliucci and Jonathan Kaplan, "The Fall and Rise of Dr Pangloss: Adaptationism and the Spandrels Paper 20 Years Later," *Trends in Ecology & Evolution* 15, no. 2 (2000): 66–70.

2. Gould, "Evolutionary Biology of Constraint," 41–42.

3. Gould and Lewontin, "The Spandrels of San Marco," 582.

4. Voltaire, *Candide: Or, Optimism* (New York: Random House, 2005), 18–19.

5. Elisabeth A. Lloyd and Stephen J. Gould, "Species Selection on Variability," *Proceedings of the National Academy of Sciences* 90, no. 2 (1993): 595–99.

6. Stephen Jay Gould, "Is a New and General Theory of Evolution Emerging?," *Paleobiology* 6, no. 1 (1980): 120.

7. Gould, 119.

8. Stephen Jay Gould, *Ever Since Darwin: Reflections in Natural History* (New York: Norton, 1977), 13. Here Gould views Darwin's theory as a prohibition against reading human morality and purpose into nature.

9. For "never say higher or lower," see Gould, 36–37. "Arborescent bush," "tiny little twig," and "fortuitous cosmic afterthought" are from Stephen Jay Gould, *Dinosaur in a Haystack: Reflections in Natural History* (New York: Harmony Books, 1995), 327.

10. Gould, *Ever Since Darwin*, 13.

11. Joel Velasco, "Why You Do What You Do; Sociobiology: A New Theory of Behavior," *Time*, 1 August 1977, 54–63.

12. Velasco, 54.

13. Velasco, 54.

14. Dorothy Nelkin and M. Susan Lindee, *The DNA Mystique: The Gene as a Cultural Icon,* 2nd ed. (Ann Arbor: University of Michigan Press, 2004), 102–26.

15. Erika L. Milam, *Looking for a Few Good Males: Female Choice in Evolutionary Biology* (Baltimore: Johns Hopkins University Press, 2010), 135–59.

16. See, e.g., David P. Barash, *Sociobiology and Behavior* (New York: Elsevier, 1977); George W. Barlow and James Silverberg, eds., *Sociobiology, beyond Nature/Nurture? Reports, Definitions, and Debate* (Boulder: Westview Press for the American Association for the Advancement of Science, 1980); James H. Hunt, ed., *Selected Readings in Sociobiology* (New York: McGraw-Hill, 1980); and King's College Sociobiology Group, ed., *Current Problems in Sociobiology* (Cambridge: Cambridge University Press, 1982).

17. Milam, *Looking for a Few Good Males,* 138–41.

18. Robert Trivers, "Parental Investment and Sexual Selection," in *Sexual Selection and the Descent of Man,* ed. B. Campbell (Chicago: Aldine, 1972), 136–79.

19. Edward O. Wilson, *On Human Nature* (Cambridge, MA: Harvard University Press, 1978), 129.

20. Wilson, 134–37.

21. Richard Dawkins, *The Selfish Gene* (Oxford: Oxford University Press, 1976).

22. Dawkins, 95.

23. Mark E. Borello, *Evolutionary Restraints: The Contentious History of Group Selection* (Chicago: University of Chicago Press, 2010), 2–6.

24. Lily E. Kay, *Who Wrote the Book of Life? A History of the Genetic Code* (Stanford, CA: Stanford University Press, 2000), 13.

25. Edward O. Wilson, *Sociobiology: The New Synthesis* (Cambridge, MA: Belknap Press of Harvard University Press, 1975), 3–6; Dawkins, *Selfish Gene,* 49–70.

26. Richard Lewontin, "Fitness, Survival and Optimality," in *Analysis of Ecological Systems,* ed. David J. Horn, Gordon Stairs, and Rodger Mitchell (Columbus: Ohio State University Press, 1979), 7–10; John Beatty, "Optimal-Design Models and the Strategy of Model Building in Evolutionary Biology," *Philosophy of Science* 47, no. 4 (1980): 534–35.

27. George F. Oster and Edward O. Wilson, *Caste and Ecology in the Social Insects* (Princeton, NJ: Princeton University Press, 1978), 295.

28. Dawkins, *Selfish Gene,* 46.

29. Trivers, "Parental Investment and Sexual Selection," 171–72; Milam, *Looking for a Few Good Males,* 135–59.

30. Gould and Lewontin, "Spandrels of San Marco," 594.

31. Stephen Jay Gould, *Ontogeny and Phylogeny* (Cambridge, MA: Harvard University Press, 1977), 38.

32. Gould, "Evolutionary Biology of Constraint," 39–52; Stephen Jay Gould, *The Mismeasure of Man,* rev. and expanded ed. (New York: Norton, 1996), 406; Gould, *I Have Landed: The End of a Beginning in Natural History* (New York: Harmony Books, 2002), 29–53.

33. Although Gould did attempt to bring these interventions together in a single unified proposal (see, e.g., Elisabeth S. Vrba and Stephen Jay Gould, "The Hierarchical Expansion of Sorting and Selection: Sorting and Selection Cannot Be Equated," *Paleobiology* 12 [1986]: 217–28; Lloyd and Gould, "Species Selection on Variability"; and Stephen Jay Gould, *The Structure of Evolutionary Theory* [Cambridge, MA: Harvard

University Press, 2002], 745–1295), there is ongoing philosophical and historical debate about whether these attempts were consistent with one another. See John Beatty, "Replaying Life's Tape," *Journal of Philosophy* 103, no. 7 (2006): 336–62; John Beatty and Eric Desjardins, "Natural Selection and History," *Biology & Philosophy* 24, no. 2 (2009): 231–46; Derek Turner, "Historical Contingency and the Explanation of Evolutionary Trends," in *Explanation in Biology: An Enquiry into the Diversity of Explanatory Patterns in the Life Sciences*, ed. Pierre-Alain Braillard and Christophe Malaterre, History, Philosophy and Theory of the Life Sciences (Dordrecht: Springer Netherlands, 2015), 73–90; and Alison K. McConwell and Adrian Currie, "Gouldian Arguments and the Sources of Contingency," *Biology & Philosophy* 32, no. 2 (2017): 243–61.

34. Stephen Jay Gould, "The Pattern of Life's History," in *The Third Culture: Beyond the Scientific Revolution*, ed. John Brockman (New York: Simon & Schuster, 1996), 54.

35. Richard C. Lewontin, "The Units of Selection," *Annual Review of Ecology and Systematics* 1 (1970): 1–18; Lewontin, *The Genetic Basis of Evolutionary Change* (New York: Columbia University Press, 1970); Lewontin, "Adaptation," *Scientific American* 239, no. 3 (1978): 212–31.

36. Gould, "Pattern of Life's History," 55.

37. Alan C. Love, "Evolutionary Morphology, Innovation, and the Synthesis of Evolutionary and Developmental Biology," *Biology & Philosophy* 18, no. 2 (2003): 309–45; Love, "Reflections on the Middle Stages of EvoDevo's Ontogeny," *Biological Theory* 1, no. 1 (2006): 94–97.

38. Gould, *Ontogeny and Phylogeny*, 7–9.

39. Gould, 8–9.

40. Gould, *The Structure of Evolutionary Theory*, 43.

41. Gould, 43.

42. Arthur J. Cain, "Introduction to General Discussion," *Proceedings of the Royal Society of London. Series B. Biological Sciences* 205 (1979): 601.

43. Ullrica Christina Olofsdotter Segerstråle, *Defenders of the Truth: The Battle for Science in the Sociobiology Debate and Beyond* (Oxford: Oxford University Press, 2000), 119–20.

44. John Maynard Smith, "Evolution: Palaeontology at the High Table," *Nature* 309, no. 5967 (1984): 401–2.

45. David Raup, "Interview with David M. Raup," in *The Paleobiological Revolution: Essays on the Growth of Modern Paleontology*, ed. David Sepkoski and Michael Ruse (Chicago: University of Chicago Press, 2009), 463.

46. David Sepkoski, *Rereading the Fossil Record: The Growth of Paleobiology as an Evolutionary Discipline* (Chicago: University of Chicago Press, 2012), 268–70.

47. Stephen Jay Gould to Thomas Schopf, 5 October 1973, Box 5, Folder 14, Thomas J. M. Schopf Papers, Smithsonian Institution Archives, Record Unit 7429.

48. Sepkoski, *Rereading the Fossil Record*, 266–67.

49. Sepkoski, 137–84. In later years the singular term *punctuated equilibrium* was more commonly used than the plural *punctuated equilibria*. I use the plural here to match Eldredge and Gould's original terminology.

50. Charles Darwin, *On the Origin of Species by Means of Natural Selection, or the Preservation of Favoured Races in the Struggle for Life* (London: John Murray, 1859), 279–311.

51. Niles Eldredge and Stephen Gould, "Punctuated Equilibria: An Alternative to Phyletic Gradualism," in *Models in Paleobiology*, ed. Thomas Schopf (San Francisco: Freeman Cooper, 1972), 82–115.

52. Stephen Jay Gould and Niles Eldredge, "Punctuated Equilibria: The Tempo and Mode of Evolution Reconsidered," *Paleobiology* 3 (1977): 115–51; Niles Eldredge, *Unfinished Synthesis: Biological Hierarchies and Modern Evolutionary Thought* (Oxford: Oxford University Press, 1985); Lloyd and Gould, "Species Selection on Variability."

53. Luis W. Alvarez, Walter Alvarez, Frank Asaro, and Helen v. Michel, "Extraterrestrial Causes for the Cretaceous-Tertiary Extinction," *Science* 208, no. 4448 (1980): 1095–1108.

54. Stephen Jay Gould, "The Paradox of the First Tier: An Agenda for Paleobiology," *Paleobiology* 11, no. 1 (1985): 2–12.

55. Stephen Jay Gould, *Time's Arrow, Time's Cycle: Myth and Metaphor in the Discovery of Geological Time* (Cambridge, MA: Harvard University Press, 1987).

56. In *Time's Arrow, Time's Cycle*, Gould drew heavily on the work of historians of the geological sciences, especially Martin Rudwick. See, e.g. Martin J. S. Rudwick, *Bursting the Limits of Time: The Reconstruction of Geohistory in the Age of Revolution* (Chicago: University of Chicago Press, 2007) and *The Meaning of Fossils: Episodes in the History of Palaeontology* (Chicago: University of Chicago Press, 2008).) However, in the book's acknowledgments, Gould claimed a kind of scholarly amnesia to excuse his inconsistent citations: "I am embarrassed that I cannot now sort out and properly attribute the bits and pieces forged together here. I am too close to the subject. I have taught the discovery of time for twenty years, and have read the three documents over and over again. . . . I simply do not remember which pieces came from my own readings of Burnet, Hutton, and Lyell, and which from Hooykaas, or Rudwick, Porter or a host of other thinkers who have inspired me" (viii).

57. Gould, *Time's Arrow, Time's Cycle*, 10–11.

58. Gould, 11.

59. Stephen Jay Gould, *Wonderful Life: The Burgess Shale and the Nature of History* (New York: Norton, 1989), 35.

60. Gould, 14.

61. Jimmy Carter to Stephen Jay Gould, 12 December 1989, Box 113, Folder 9, Gould Papers.

62. The debate between Simon Conway Morris and Gould is the most famous instance of *Wonderful Life*'s entry into debates over the role of chance in the origin of life and its implications for divine creation. See Simon Conway Morris, *The Crucible of Creation: The Burgess Shale and the Rise of Animals* (Oxford: Oxford University Press, 1998); and Morris, *Life's Solution: Inevitable Humans in a Lonely Universe* (Cambridge: Cambridge University Press, 2003). For further discussion, see Beatty, "Replaying Life's Tape"; and Alison K. McConwell, "Walking the Line: A Tempered View of Contingency and Convergence in Life's History," *Acta Biotheoretica* 67, no. 3 (2019): 253–64.

63. Stephen Jay Gould to Jimmy Carter, 21 December 1989, Gould Papers.

64. For example, in 1984 one of the primary figures of the modern evolutionary synthesis, Ledyard Stebbins, wrote an impassioned letter to Gould accusing him of using his *Natural History* column to "take another slap at so-called fuddy-duddies like

myself." Ledyard Stebbins to Stephen Jay Gould, 20 July 1984, Box 698, Folder 2, Gould Papers.

65. Stephen Jay Gould, *Full House: The Spread of Excellence from Plato to Darwin* (Cambridge, MA: Harvard University Press, 1996), 18.

66. Gould, 4.

67. Donald Symons, *The Evolution of Human Sexuality* (New York: Oxford University Press, 1979), 27–28.

68. Symons, 27.

69. Jerome H. Barkow, Leda Cosmides, and John Tooby, eds., *The Adapted Mind: Evolutionary Psychology and the Generation of Culture* (New York: Oxford University Press, 1991), 5.

70. Rachel Louise Snyder, *No Visible Bruises: What We Don't Know About Domestic Violence Can Kill Us* (New York: Bloomsbury, 2019), 11–15; Aya Gruber, *The Feminist War on Crime: The Unexpected Role of Women's Liberation in Mass Incarceration* (Berkeley: University of California Press, 2020), 1–18.

71. Randy Thornhill and Craig T. Palmer, *A Natural History of Rape: Biological Bases of Sexual Coercion* (Cambridge, MA: MIT Press, 2000), 4–9.

72. Cheryl Brown Travis, *Evolution, Gender, and Rape* (Cambridge, MA: MIT Press, 2003).

73. Wilson, *Sociobiology*, 555.

74. Christopher Leslie Ryan and Cacilda Jetha, *Sex at Dawn: How We Mate, Why We Stray, and What It Means for Modern Relationships* (New York: HarperCollins, 2010), 103.

75. Matt Ridley, *The Red Queen: Sex and the Evolution of Human Nature* (New York: Penguin Books, 1993), 181.

76. Steven Pinker, *The Blank Slate: The Modern Denial of Human Nature* (New York: Penguin Books, 2003), 337–43. *The Blank Slate* was originally published by Viking in 2002, but all references are to the Penguin paperback edition of 2003.

77. Pinker, 356.

78. Pinker, 352.

79. Pinker, 345.

80. Pinker, 328.

81. Pinker, 329.

82. Pinker, 296.

83. Dawkins, *Selfish Gene*, 3.

84. Martha McCaughey, *The Caveman Mystique: Pop-Darwinism and the Debates Over Sex, Violence, and Science* (New York: Routledge, 2008), 1–20.

85. Evgeny Morozov, "Jeffrey Epstein's Intellectual Enabler," *New Republic*, 22 August 2019, https://newrepublic.com/article/154826/jeffrey-epsteins-intellectual-enabler.

86. Daniel K. Williams, *God's Own Party: The Making of the Christian Right* (Oxford: Oxford University Press, 2012), 105–32.

87. Letha Scanzoni and Nancy Hardesty, *All We're Meant to Be: A Biblical Approach to Women's Liberation* (Waco, TX: Word Books, 1974).

88. Council on Biblical Manhood and Womanhood, "The Danvers Statement" [December 1987], 26 June 2007, https://cbmw.org/uncategorized/the-danvers-statement/.

89. Council on Biblical Manhood and Womanhood, "Nashville Statement, Article 10" [August 2017], https://cbmw.org/nashville-statement.

90. Gregg Johnson, "The Biological Basis for Gender-Specific Behavior," in *Recovering Biblical Manhood and Womanhood: A Response to Evangelical Feminism*, ed. John Piper and Wayne Grudem (Wheaton, IL: Crossway Books, 1991), 281.

91. Martin Daly and Margo Wilson, *Sex, Evolution, and Behavior: Adaptations for Reproduction* (Boston: Willard Grant, 1978); Myrna Perez Sheldon, "Wild at Heart: How Sociobiology and Evolutionary Psychology Helped Influence the Construction of Heterosexual Masculinity in American Evangelicalism," *Signs: Journal of Women in Culture and Society* 42, no. 4 (2017): 983.

92. Sheldon, "Wild at Heart," 987–93.

93. W. E. B. Du Bois, "The Negro Scientist," *American Scholar* 8, no. 3 (1939): 309–20.

94. Du Bois, 310.

95. Daniel Nolan, "Why Historians (and Everyone Else) Should Care about Counterfactuals," *Philosophical Studies* 163, no. 2 (2013): 333–34.

96. John Beatty and Isabel Carrera, "When What Had to Happen Was Not Bound to Happen: History, Chance, Narrative, Evolution," *Journal of the Philosophy of History* 5, no. 3 (2011): 471, 481–91.

97. Du Bois, "Negro Scientist," 318.

98. Gould, *Wonderful Life*, 291.

99. Two particularly notable biological studies are Zachary D. Blount, Richard E. Lenski, and Jonathan B. Losos, "Contingency and Determinism in Evolution: Replaying Life's Tape," *Science* 362, no. 6415 (2018): eaam5979, https://doi.org/10.1126/science.aam5979; and Michael Travisano, Judith A. Mongold, Albert F. Bennett, and Richard E. Lenski, "Experimental Tests of the Roles of Adaptation, Chance, and History in Evolution," *Science* 267, no. 5194 (1995): 87–90. For further discussion of these studies, see, e.g., Paul H. Harvey and Linda Partridge, "Different Routes to Similar Ends," *Nature* 392, no. 6676 (1998): 552–53; S. B. Emerson, "A Macroevolutionary Study of Historical Contingency in the Fanged Frogs of Southeast Asia," *Biological Journal of the Linnean Society* 73, no. 1 (2001): 139–51; and R. Craig MacLean and Graham Bell, "Divergent Evolution during an Experimental Adaptive Radiation," *Proceedings of the Royal Society of London. Series B. Biological Sciences* 270, no. 1524 (2003): 1645–50.

100. David Sepkoski, "'Replaying Life's Tape': Simulations, Metaphors, and Historicity in Stephen Jay Gould's View of Life," *Studies in History and Philosophy of Science Part C: Studies in History and Philosophy of Biological and Biomedical Sciences* 58 (2016): 73–81.

101. Beatty, "Replaying Life's Tape."

102. Douglas H. Erwin, "Was the Ediacaran-Cambrian Radiation a Unique Evolutionary Event?," *Paleobiology* 41, no. 1 (2015): 1–15; Erwin, "Wonderful Life Revisited: Chance and Contingency in the Ediacaran-Cambrian Radiation," in *Chance in Evolution*, ed. Grant Ramsey and Charles H. Pence (Chicago: University of Chicago Press, 2016), 277–98.

103. "The Mismeasure of Man, publicity," Box 354, Folder 6, Gould Papers; "Wonderful Life, publicity," Box 267, Folder 9, Gould Papers.

104. Here Gould does not reference sociobiology directly but discusses his view that

iconographies of predictable progress (i.e. those created by a conviction in the optimizing force of selection) create hierarchies among living things as well as human races. Gould, *Wonderful Life*, 28.

Chapter 4 • Scientists Confront Creationism

Epigraph: Stephen Jay Gould, "A Visit to Dayton," *Natural History* 90 (1981): 17.

1. Gould, 12, 14.

2. McLean v. Arkansas Bd. of Ed., 529 F. Supp. 1255 (E.D. Ark. 1982).

3. Marcel Chotkowski LaFollette, introduction to *Creationism, Science, and the Law: The Arkansas Case, ed.* LaFollette (Cambridge, MA: MIT Press, 1983), 1–2; Edward J. Larson, *Trial and Error: The American Controversy Over Creation and Evolution*, 3rd ed. (New York: Oxford University Press, 2003), 162–66.

4. Stephen Jay Gould, "Evolution as Fact and Theory," *Discover*, May 1981, 34.

5. For historical accounts in this vein, see, e.g., Edward J. Larson, *Trial and Error: The American Controversy over Creation and Evolution* (New York: Oxford University Press, 1985); and Ronald L. Numbers, *The Creationists: The Evolution of Scientific Creationism* (New York: Knopf , 1992). The journalistic coverage of creationism was extensive. For characterizations of this, see Jeffrey P Moran, *American Genesis: The Antievolution Controversies from Scopes to Creation Science* (New York: Oxford University Press, 2012).

6. In his review of the "legends" of the 1925 Scopes trial, Ronald Numbers stresses the historiographic disagreements that persisted over the trial as a reflection of the cultural importance the trial holds for historians and the American public alike. See Ronald L. Numbers, *Darwinism Comes to America* (Cambridge, MA: Harvard University Press, 1998), 76–91. Recent work has stressed that evangelical religious communities sought political engagement throughout the twentieth century. See esp. Darren Dochuk, *From Bible Belt to Sunbelt: Plain-Folk Religion, Grassroots Politics, and the Rise of Evangelical Conservatism* (New York: Norton, 2010); and Matthew Avery Sutton, *American Apocalypse: A History of Modern Evangelicalism* (Cambridge, MA: Belknap Press of Harvard University Press, 2014).

7. John L. Rudolph, *Scientists in the Classroom: The Cold War Reconstruction of American Science Education* (New York: Palgrave Macmillan, 2002); Jeffrey P. Moran, *Teaching Sex: The Shaping of Adolescence in the 20th Century* (Cambridge, MA: Harvard University Press, 2000); Jamie Cohen-Cole, *The Open Mind: Cold War Politics and the Sciences of Human Nature* (Chicago: University of Chicago Press, 2014), 217–53.

8. Matthew Sutton argues that this moral transformation occurred earlier in the twentieth century with the emergence of the fundamentalist movement. Sutton, *American Apocalypse*, 6. Here I note the continuation of this transformation with the emergence of the partisan politics of the Christian Right leading up to and after the 1980 presidential election.

9. Stephen Jay Gould, "Evolution Wins Again," *New York Times*, 12 January 1982, A15.

10. Gould; Stephen Jay Gould, "Born Again Creationism," *Science for the People* 13, no. 5 (1981): 10–11, 37; Gould, "Evolution as Fact and Theory"; Gould, "Moon, Mann, and Otto," *Natural History* 91, no. 3 (1983): 4–10.

11. Daniel T. Rodgers, *Age of Fracture* (Cambridge, MA: Harvard University Press,

2011), 165–78; Andrew Hartman, *A War for the Soul of America: A History of the Culture Wars* (Chicago: University of Chicago Press, 2015), 70–101.

12. Sutton, *Apocalypse*, x.

13. Adam R. Shapiro, *Trying Biology: The Scopes Trial, Textbooks, and the Antievolution Movement in American Schools (Chicago: University of Chicago Press, 2013), 78–79.*

14. Paul A. Carter, "The Fundamentalist Defense of the Faith," in *Change and Continuity in Twentieth Century America: The 1920's*, ed John Braeman, Robert H. Bremner, and David Brody (Columbus: Ohio State University Press, 1968), 204–5; Margaret Lamberts Bendroth, *Fundamentalists in the City: Conflict and Division in Boston's Churches, 1885–1950* (Oxford: Oxford University Press, 2005).

15. Dochuk, *From Bible Belt to Sunbelt*, 4–8.

16. T. M. Luhrmann, *When God Talks Back: Understanding the American Evangelical Relationship with God* (New York: Random House, 2012), xv.

17. Thomas A. Fudge, *Christianity without the Cross: A History of Salvation in Oneness Pentecostalism* (Parkland, FL: Universal, 2003), 2–5; Allan Anderson, *An Introduction to Pentecostalism: Global Charismatic Christianity* (Cambridge: Cambridge University Press, 2013).

18. Daniel K. Williams, *God's Own Party: The Making of the Christian Right* (Oxford: Oxford University Press, 2012), 193–94.

19. Frank Schaeffer, *Crazy for God: How I Grew Up as One of the Elect, Helped Found the Religious Right, and Lived to Take All (or Almost All) of It Back* (New York: Carroll & Graf, 2007); Barry Hankins, *Francis Schaeffer and the Shaping of Evangelical America* (Grand Rapids, MI: Eerdmans, 2008). Francis (Frank) Schaeffer, son of Francis Schaeffer, interviews by author, January 2016, August 2018, June 2022.

20. The series *Whatever Happened to the Human Race?* was produced by Vision Video, in Worcester, Pennsylvania, in 1979.

21. C. Everett Koop and Francis A. Schaeffer, *Whatever Happened to the Human Race?* (Westchester, IL: Crossway Books, 1984), 14.

22. Williams, *God's Own Party*, 159–86.

23. Koop and Schaeffer, *Whatever Happened to the Human Race?*, 4.

24. Koop and Schaeffer, 9.

25. Edward O. Wilson, *Sociobiology: The New Synthesis* (Cambridge, MA: Belknap Press of Harvard University Press, 1975), 562, quoted in Koop and Schaeffer, *Whatever Happened to the Human Race?*, 9–10.

26. Koop and Schaeffer, *Whatever Happened to the Human Race?*, 10.

27. Christine Rosen, *Preaching Eugenics: Religious Leaders and the American Eugenics Movement* (Oxford: Oxford University Press, 2004), 4; Sharon Mara Leon, *An Image of God: The Catholic Struggle with Eugenics* (Chicago: University of Chicago Press, 2013); Williams, *God's Own Party*, 116.

28. Williams, *God's Own Party*, 154–55; Daniel K. Williams, *Defenders of the Unborn: The Pro-Life Movement before Roe v. Wade* (New York: Oxford University Press, 2016), 237.

29. Jerry Falwell, *Listen, America!* (New York: Doubleday, 1980), 253–54.

30. Henry M. Morris and John Clement Whitcomb, *The Genesis Flood: The Biblical Record and Its Scientific Implications* (Philadelphia: Presbyterian and Reformed Publishing, 1961).

31. Documents, including Bible study curriculum, *Acts and Facts* issues, in author's personal manuscript collection.

32. Concerned Women for America, "Our History," accessed 1 April 2019, https://concerned women.org/about/our-history/; Ronnee Schreiber, *Righting Feminism: Conservative Women and American Politics* (Oxford : Oxford University Press, 2008).

33. Balanced Treatment for Creation-Science and Evolution-Science Act, Pub. L. No. 590 (1981).

34. Roger Lewin, "A Response to Creationism Evolves," *Science* 214, no. 4521 (1981): 635–36.

35. Lewin, 635.

36. Lewin and Leakey coauthored a number of books, including Richard E. Leakey and Roger Lewin, *Origins: What New Discoveries Reveal about the Emergence of Our Species and Its Possible Future* (New York: Dutton, 1977); Leakey and Lewin, *People of the Lake: Mankind and Its Beginnings* (New York: Avon, 1979); and Leakey and Lewin, *Origins Reconsidered: In Search of What Makes Us Human* (New York: Doubleday, 1992).

37. Roger Lewin, "Evolutionary Theory under Fire," *Science* 210, no. 4472 (1980): 883–87.

38. Lewin, "Response to Creationism Evolves," 635.

39. Ronald Reagan, quoted in Constance Holden, "Republican Candidate Picks Fight with Darwin," *Science* 209, no. 4462 (1980): 1214.

40. Stephen Jay Gould to Helena Curtis, 14 November 1980, Box 108, Folder 12, Gould Papers; Stephen Jay Gould to Boyce Rensberger, 14 November 1980, Box 10, Folder 12, Gould Papers.

41. Francisco Ayala to SSE Education Committee, 29 December 1980, Series III: Officer's Correspondence, SSE Records, American Philosophical Society, Philadelphia.

42. Ayala would become president of the American Association for the Advancement of Science in 1995. He subsequently resigned from his professorship at the University of California, Irvine, after a report substantiated sexual harassment claims against him. Meredith Wadman, "Report Gives Details of Sexual Harassment Allegations That Felled a Famed Geneticist," *Science*, 20 July 2018, https://www.science.org/content/article/report-gives-details-sexual-harassment-allegations-felled-famed-geneticist.

43. Audra J. Wolfe, *Freedom's Laboratory: The Cold War Struggle for the Soul of Science* (Baltimore: Johns Hopkins University Press, 2018), 138–42.

44. "Obituary of John A. Moore," *National Center for Science Education*, accessed 29 February 2012, http://ncse.com/ncser/22/6/john-moore-champion-evolution.

45. Ayala to SSE Education Committee, 29 December 1980.

46. Minutes from SSE annual meeting, University of Iowa, 28 June 1981, Series III: Secretary's Correspondence, SSE Records, American Philosophical Society, Philadelphia.

47. Vassiliki Betty Smocovitis, "Disciplining Evolutionary Biology: Ernst Mayr and the Founding of the Society for the Study of Evolution and *Evolution* (1939–1950)," *Evolution* 48, no. 1 (1994): 1–8.

48. The difficulties that professional science organizations had with direct political action are discussed in Kelly Moore, *Disrupting Science: Social Movements, American Scientists, and the Politics of the Military, 1945–1975* (Princeton, NJ: Princeton University Press, 2009), 1–7.

49. Stephen Jay Gould to Eric Christianson, 15 July 1980, Box 108, Folder 8, Gould Papers.

50. Stephen Jay Gould to "Friends," 24 November 1981, Box 108, Folder 24, Gould Papers.

51. Gould made this point in several places, but most clearly in Stephen Jay Gould, "Creationism: Genesis vs. Geology," *Atlantic*, 1 September 1982, 10–17, reprinted in Ashley Montagu, ed., *Science and Creationism* (Oxford: Oxford University Press, 1984), 126–35.

52. Lewin, "Response to Creationism Evolves," 635.

53. Gould to "Friends," 24 November 1981, Gould Papers.

54. "Deposition of Stephen Jay Gould, 27 November 1981," McLean v. Arkansas Documentation Project, 23–24, http://www.antievolution.org/projects/mclean/new_site/depos/pf_gould_dep.htm.

55. Gould, "Moon, Mann, and Otto," reprinted in Gould, *Hen's Teeth and Horse's Toes: Further Reflections in Natural History* (New York: Norton, 1983), 280–90.

56. The work of all three is cited in the reprint of the essay in Gould, *Hen's Teeth and Horse's Toes*, 282.

57. Judith V. Grabiner and Peter D. Miller, "Effects of the Scopes Trial," *Science* 185, no. 4154 (1974): 833, 836.

58. Dorothy Nelkin, *Science Textbook Controversies and the Politics of Equal Time* (Cambridge, MA: MIT Press, 1977).

59. Gould, "Visit to Dayton," 17.

60. McLean v. Arkansas Bd. Of Ed., 529 F. Supp. 1255 (E.D. Ark. 1982) https://law.justia.com/cases/federal/district-courts/FSupp/529/1255/2354824/.

61. Larson, *Trial and Error*, 191, 164–65; William R. Overton, "*McLean v. Arkansas* Opinion, U.S. District Judge, Eastern District of Arkansas, Western Division," in *Creationism, Science, and the Law: The Arkansas Case*, ed. Marcel Chotkowski LaFollette (1983; reprint, Cambridge, MA: MIT Press, 1984), 60.

62. Overton, "*McLean v. Arkansas* Opinion," 9.

63. Overton, 60.

64. Mark E. Herlihy, "Scientific Disputes and Legal Strategies," in LaFollette, *Creationism, Science, and the Law* (1984), 99.

65. Gould, "Evolution Wins Again"; reprint of Clarence Darrow, "Evolution A 'Crime,'" and H. L. Mencken, "Scopes: Infidel," *New York Times, 12* January 1982, A15.

66. Gould, "Evolution Wins Again."

67. For example, Gould wrote to Henry L. Gates Jr. in this period, "I know that many of my colleagues court publishers and know about what they do. I tend to specialize in avoiding such knowledge and contact." Stephen Jay Gould to Henry L. Gates Jr., 5 August 1980, Box 108, Folder 9, Gould Papers.

68. This may have been partly owing to Gould's suffering from cancer a few months after the trial concluded. He wrote less in the period immediately after the trial than he presumably would have done otherwise, and this perhaps was one reason for the multiple uses of this particular piece.

69. Gould, "Evolution as Fact and Theory," 35.

70. Herbert F. Vetter, *Speak Out Against the New Right* (Boston: Beacon, 1982), 1.

71. Herbert F. Vetter, *Speak Out Against the New Right*, 2nd ed. (Cambridge, MA: Harvard Square Library, 2004), 9.

72. Herbert Vetter, interview by author, 18 April 2012, Cambridge, MA.

73. Materials related to the Arkansas trial, Box 2, Folders 4–6, Stanley L. Weinberg Papers, MS 640, Special Collections Department, Iowa State University Library, Ames.

74. Lewin, "Response to Creationism Evolves," 636.

75. John H. Maudlin to Stanley Weinberg, 13 December 1981, Box 10, Folder 9, Weinberg Papers.

76. "Memorandum to Liaisons for Committees of Correspondence," January 1981, Box 12, Folder 7, Weinberg Papers.

77. Stanley Weinberg to Stephen Jay Gould, 23 May 1978 and 7 August 1979, Box 12, Folder 8, Gould Papers.

78. "Deposition of Stephen Jay Gould," 15.

79. Eugenie Scott to Stephen Jay Gould, 18 February 1988, Box 93, Folder 17, Gould Papers.

80. Scott to Gould, 18 February 1988.

81. Gould, "Visit to Dayton," 8–22.

82. Rachel Martin, "Feeling Left Behind, White Working-Class Voters Turned Out For Trump," *NPR*, National, 13 November 2016,https://www.npr.org/2016/11/13/501904167/feeling-left-behind-white-working-class-voters-turned-out-for-trump; Emma Green, "It Was Cultural Anxiety That Drove White, Working-Class Voters to Trump," *Atlantic*, 9 May 2017, https://www.theatlantic.com/politics/archive/2017/05/white-working-class-trump-cultural-anxiety/525771/. Critiques of the disenfranchised working-class narrative of the 2016 election include, for example, Sarah Jaffe, "Whose Class Is It Anyway? The 'White Working Class' and the Myth of Trump," in *Labor in the Time of Trump*, ed. Jasmine Kerrissey, Eve Weinbaum, Clare Hammonds, Tom Juravich, and Dan Clawson (Ithaca, NY: ILR Press, 2020), 87–105; and Nicholas Carnes and Noam Lupu, "The White Working Class and the 2016 Election," *Perspectives on Politics* 19, no. 1 (2021): 55–72.

83. Gould, "Visit to Dayton," 12–14.

84. Richard Dawkins, *The Blind Watchmaker: Why the Evidence of Evolution Reveals a Universe Without Design* (New York: Norton, 1986), 251.

85. For examinations of evangelicalism, class, and race, see Bethany Moreton, *To Serve God and Wal-Mart: The Making of Christian Free Enterprise* (Cambridge, MA: Harvard University Press, 2009); Nadine Hubbs, *Rednecks, Queers, and Country Music* (Berkeley: University of California Press, 2014); Kevin Michael Kruse, *One Nation under God: How Corporate America Invented Christian America* (Princeton, NJ: Princeton University Press, 2015); and Janelle S. Wong, *Immigrants, Evangelicals, and Politics in an Era of Demographic Change* (New York: Russell Sage Foundation, 2018).

86. Larson, *Trial and Error*, 28–29; Ronald L. Numbers, *The Creationists: From Scientific Creationism to Intelligent Design*, expanded ed. (Cambridge, MA: Harvard University Press, 2006), 29, 108; Constance A. Clark, *God—or Gorilla: Images of Evolution in the Jazz Age* (Baltimore: Johns Hopkins University Press, 2008), 173–75; Shapiro, *Trying Biology*, 15–16.

87. Larson, *Trial and Error*, 49.

88. Several scholars in rhetoric and communication studies have suggested a populist framework for analyzing creationism: Michael W. Apple, "Bringing the World to God: Education and the Politics of Authoritarian Religious Populism," *Discourse:*

Studies in the Cultural Politics of Education 22, no. 2 (2001): 149–72; Taner Edis, "A Revolt Against Expertise: Pseudoscience, Right-Wing Populism, and Post-Truth Politics," *Disputatio. Philosophical Research Bulletin* 9, no. 13 (2020): 67–95; Hande Eslen-Ziya, "Knowledge, Counter-Knowledge, Pseudo-Science in Populism," in *Populism and Science in Europe*, ed. Eslen-Ziya and Alberta Giorgi, Palgrave Studies in European Political Sociology (Cham, Switzerland: Springer International, 2022), 25–41.

89. Numbers, *Creationists*, 312–16.

90. Susan L. Trollinger and William Vance Trollinger Jr., *Righting America at the Creation Museum*, Medicine, Science, and Religion in Historical Context (Baltimore: Johns Hopkins University Press, 2016).

91. Boston Women's Health Book Collective, *Our Bodies, Ourselves: A Book by and for Women* (Boston: Boston Women's Health Collective, 1970); Sandra Morgen, *Into Our Own Hands: The Women's Health Movement in the United States, 1969–1990* (New Brunswick, NJ: Rutgers University Press, 2002), 16–40.

92. Alondra Nelson, *Body and Soul: The Black Panther Party and the Fight against Medical Discrimination* (Minneapolis: University of Minnesota Press, 2011), 75–78.

93. Darrel Enck-Wanzer, *The Young Lords: A Reader* (New York: New York University Press, 2010), 200–201.

94. Stuart W. Leslie, *The Cold War and American Science: The Military-Industrial-Academic Complex at MIT and Stanford* (New York: Columbia University Press, 1993); Rebecca S. Lowen, *Creating the Cold War University: The Transformation of Stanford* (Berkeley: University of California Press, 1997); Audra J. Wolfe, *Competing with the Soviets: Science, Technology, and the State in Cold War America* (Baltimore: Johns Hopkins University Press, 2013), 9–10.

95. Michael E. Latham, *Modernization as Ideology: American Social Science and "Nation Building" in the Kennedy Era* (Chapel Hill: University of North Carolina Press, 2000); Wolfe, *Competing with the Soviets*, 55–73.

96. Elaine Tyler May, *Homeward Bound: American Families in the Cold War Era* (New York: Basic Books, 2008), 19–23.

Chapter 5 • The Accidental Creationists

Epigraph: Stephen Jay Gould, "Evolution as Fact and Theory," *Discover Magazine*, May 1981, 37.

1. Ronald L. Numbers, *The Creationists: From Scientific Creationism to Intelligent Design*, 2nd ed. (Berkeley: University of California Press, 2006), 387.

2. Stephen Jay Gould, "Dorothy, It's Really Oz," *Time*, 23 August 1999, https://content.time.com/time/subscriber/article/0,33009,991791-2,00.html.

3. Numbers, *Creationists*, 387–88.

4. Phillip E. Johnson, *Darwin on Trial* (Downers Grove, IL: InterVarsity, 1991); Michael J. Behe, *Darwin's Black Box: The Biochemical Challenge to Evolution* (New York: Simon & Schuster, 1996).

5. Kitzmiller v. Dover Area School Dist., 400 F. Supp. 2d 707 (M.D. Pa. 2005).

6. Gould, "Dorothy, It's Really Oz." For a discussion of creationism's global presence, see Numbers, *Creationists*, 399–431.

7. Robert Wright, "The Accidental Creationist: Why Stephen Jay Gould Is Bad for Evolution," *New Yorker*, 13 December 1999, 56.

8. Wright, 56.

9. Richard Dawkins, *The Blind Watchmaker: Why the Evidence of Evolution Reveals a Universe Without Design* (New York: Norton, 1986). Throughout this period, authors used both *equilibria* and *equilibrium* when discussing Eldrege and Gould's theory; Dawkins used both in *The Blind Watchmaker.* I use *equilibria* throughout this chapter to match Eldredge and Gould's terminology and to avoid confusion.

10. Dawkins, 251.

11. Daniel Clement Dennett, *Darwin's Dangerous Idea: Evolution and the Meanings of Life* (New York: Simon & Schuster, 1995), 262–312.

12. Paul R. Gross and Norman Levitt, *Higher Superstition: The Academic Left and Its Quarrels with Science,* 2nd ed. (Baltimore: Johns Hopkins University Press, 1998), 9, 60.

13. Niles Eldredge and Stephen Gould, "Punctuated Equilibria: An Alternative to Phyletic Gradualism," in *Models in Paleobiology,* ed. Thomas Schopf (San Francisco: Freeman Cooper, 1972), 82–115.

14. Henry M. Morris, *Scientific Creationism,* 2nd ed. (El Cajon, CA: New Leaf Publishing Group, 1985), viii.

15. Stephen Jay Gould, "Evolution's Erratic Pace," *Natural History* 86, no. 5 (1977): 12–16, reprinted as "The Episodic Nature of Evolutionary Change," in *The Panda's Thumb: More Reflections in Natural History* (New York: Norton, 1980), 179–85.

16. Thomas H. Huxley to Charles Darwin, 23 November 1859, , quoted in Gould, "Evolution's Erratic Pace," 12.

17. Gould, "Evolution's Erratic Pace," 12.

18. Huxley to Darwin, 23 November 1859, quoted in Gould, 12.

19. Gould, 14.

20. Although few archival records of the meeting exist, the conference lived on in various professional memories, magnified by the press the conference received. David Sepkoski, *Rereading the Fossil Record: The Growth of Paleobiology as an Evolutionary Discipline* (Chicago: University of Chicago Press, 2012), 363–71. In Gould's correspondence before and after the conference he insisted that he was not responsible for organizing the meeting. For example, in a letter to Eric Charmov he wrote, "I am going to attend the Macroevolution Conference in Chicago, but John is wrong in thinking that I have played any part in organizing it." Stephen Jay Gould to Eric L. Charmov, 8 October 1980, Box 108, Folder 11, Gould Papers.

21. Roger Lewin, "Evolutionary Theory under Fire," *Science* 210, no. 4472 (1980): 883–87; Jerry Adler and John Carey, "Is Man a Subtle Accident?," *Newsweek,* 3 November 1980; Boyce Rensberger, "Recent Studies Spark Revolution in Interpretation of Evolution," *New York Times,* 4 November 1980.

22. Stephen Jay Gould, *The Structure of Evolutionary Theory* (Cambridge, MA: Harvard University Press, 2002), 982.

23. Gould, 983.

24. Gould, 982.

25. Stephen Jay Gould and Niles Eldredge to Leon Jaroff, 7 October 1980, Box 108, Folder 11, Gould Papers.

26. Dennis Hevesi, "Leon Jaroff, Editor at Time and Discover Magazines, Dies at 85," *New York Times,* 21 October 2012.

27. Leon Jaroff to Stephen Jay Gould, 13 December 1978, Box 5, Folder 39, Gould Papers.

28. James Gorman, "The History of a Theory," *New York Times*, 20 November 1977, sec. BR, 4; Gorman, interview by author, 18 December 2012, Cambridge, MA.

29. James Gorman to Stephen Jay Gould, 16 September 1980, Box 18, Folder 20, Gould Papers.

30. Gorman interview.

31. James Gorman, "Stephen Jay Gould and the Fight Against Creationism," *Discover*, January 1982, 56.

32. Jerry Adler, interview by author, 11 June 2012, Cambridge, MA.

33. Loose tracts, dated 1985–95, Institute for Creation Research, San Diego, CA, collected 2 June 2011, in author's collection.

34. Johnson, *Darwin on Trial*, 60–62, 184–85; Michael Denton, *Evolution: A Theory in Crisis* (Bethesda, MD: Adler & Adler, 1986), 192–95.

35. Kim Sterelny, *Dawkins vs. Gould: Survival of the Fittest*, Revolutions in Science (New York: Icon Books, 2001); Elliot Sober, *Philosophy of Biology*, 2nd ed., Dimensions of Philosophy Series (Boulder, CO: Westview Press, 2000), 88–142.

36. Andrew Brown, *The Darwin Wars: The Scientific Battle for the Soul of Man* (London: Simon & Schuster, 1999); Richard Morris, *The Evolutionists: The Struggle for Darwin's Soul* (New York: W. H. Freeman, 2001); Michael Ruse, *The Evolution Wars: A Guide to the Debates* (New Brunswick, NJ: Rutgers University Press, 2001).

37. Cover of *Discover*, February 1982; James Gorman, "Judgment Day for Creationism," *Discover*, February 1982, 14–18.

38. Constance A. Clark, *God—or Gorilla: Images of Evolution in the Jazz Age* (Baltimore: Johns Hopkins University Press, 2008); Clark, "'You Are Here': Missing Links, Chains of Being, and the Language of Cartoons," *Isis* 100, no. 3 (2009): 571–89; Janet Browne, "Looking at Darwin: Portraits and the Making of an Icon," *Isis* 100, no. 3 (2009): 542–70; Nadine Weidman, "Popularizing the Ancestry of Man: Robert Ardrey and the Killer Instinct," *Isis* 102, no. 2 (2011): 269–99; Lukas Rieppel, "Bringing Dinosaurs Back to Life: Exhibiting Prehistory at the American Museum of Natural History," *Isis* 103, no. 3 (2012): 460–90.

39. Stephen Jay Gould to Paul Kurtz, 21 December 1981, Box 109, Folder 1, Gould Papers.

40. Paul Kurtz, *In Defense of Secular Humanism* (Buffalo, NY: Prometheus Books, 1983).

41. The period 1875–1925 has been described as the "eclipse of Darwinism" to highlight the competition natural selection faced from other explanations for evolutionary change. Peter J. Bowler, *The Eclipse of Darwinism: Anti-Darwinian Evolution Theories in the Decades around 1900* (Baltimore: Johns Hopkins University Press, 1983). Julian Huxley described the period in this way in his survey of evolutionary theory before the triumph of the "evolutionary synthesis." Julian Huxley, *Evolution: The Modern Synthesis* (London: Allen & Unwin, 1942), 22–28. Casting this period as an "eclipse" was critical to Huxley's advocacy for modern evolutionary synthesis.

42. The models referred to here were developed by R. A. Fisher, J. B. S. Haldane, and Sewall Wright. See R. A. Fisher "XV.—The Correlation between Relatives on the Supposition of Mendelian Inheritance," *Transactions of the Royal Society of Edinburgh* 52

(1919): 399–433; Fisher, *The Genetical Theory of Natural Selection* (Oxford: Oxford University Press, 1930); J. B. S. Haldane, *The Causes of Evolution* (London: Longmans, 1932); and Sewall Wright, "Statistical Genetics and Evolution," *Bulletin of the American Mathematical Society* 48 (1942): 223–46. For a summary of how these models came together to found population genetics, see Edward Larson, *Evolution: The Remarkable History of a Scientific Theory* (New York: Modern Library, 2004), 151–74.

43. Theodosius Dobzhansky, "Nothing in Biology Makes Sense Except in the Light of Evolution," *American Biology Teacher* 35 (1973): 125.

44. Vassiliki Betty Smocovitis, "Disciplining Evolutionary Biology: Ernst Mayr and the Founding of the Society for the Study of Evolution and *Evolution* (1939–1950)," *Evolution* 48, no. 1 (1994): 1–8.

45. Vassiliki Betty Smocovitis, "The 1959 Darwin Centennial Celebration in America," *Osiris* 14 (1999): 274–323.

46. Erika Lorraine Milam, "The Equally Wonderful Field: Ernst Mayr and Organismic Biology," *Historical Studies in the Natural Sciences* 40, no. 3 (2010): 279–317.

47. Gould helped Mayr with the questionnaires that were circulated to the architects of the synthesis and other participants. Ernst Mayr to Stephen Jay Gould, 1 August 1974, Box 815, Folder 2, Gould Papers.

48. Edward O. Wilson, *Naturalist* (Washington, DC: Island Press, 1994), 286.

49. Wilson, 232.

50. Richard Lewontin, *The Genetic Basis of Evolutionary Change* (New York: Columbia University Press, 1970).

51. Stephen J. Gould, "Is a New and General Theory of Evolution Emerging?," *Paleobiology* 6, no. 1 (1980): 120.

52. Edwin Barber, "This View of Stephen Jay Gould," *Natural History* 108, no. 9 (2002): 48.

53. Stephen Jay Gould, *The Mismeasure of Man* (New York: Norton, 1981), 321–22.

54. Stephen Jay Gould, *Ever Since Darwin: Reflections in Natural History* (New York: Norton, 1977), 45.

55. Gould, 45. Here Gould was emphasizing what he read in Darwin as a stress on adaptation to immediate environments rather than "intrinsic trends toward higher states." He pointed to Darwin's discussion of degenerates (i.e., "anatomical simplification in parasites") as proof.

56. Stephen Jay Gould, *Full House: The Spread of Excellence from Plato to Darwin* (Cambridge, MA: Harvard University Press, 2011), 141.

57. Gould, 140.

58. Smocovitis, "1959 Darwin Centennial Celebration in America," 317; David Kohn, *The Darwinian Heritage* (Princeton, NJ: Princeton University Press, 1986), 1–5.

59. On Gould's relationship to the historical work of Darwin studies, see Frederick Burkhardt to Stephen Jay Gould, 1995, Box 2, Folder 8, Frederick Burkhardt Papers, 1983–2009, Division of Rare and Manuscript Collections, 4727, Cornell University Library, Ithaca, NY; Burkhardt to Gould, 1997, Box 2, Folder 8, Burkhardt Papers; Stephen Jay Gould, foreword to Frederick Burkhardt, *Charles Darwin's Letters: A Selection, 1825–1859* (Cambridge: Cambridge University Press, 1996), ix–xx; and Gould, "Why Darwin? Review of J. Browne, Voyaging: The Life of Charles Darwin," *New York Review of Books* 43 (1996): 10–14, reprinted as "A Sly Dullard Named Darwin: Recogniz-

ing the Multiple Facets of Genius," in Stephen Jay Gould, *The Lying Stones of Marrakech: Penultimate Reflections in Natural History* (Cambridge, MA: Belknap Press of Harvard University Press, 2011), 169–81. For historical biographies of Darwin in this vein, see esp. E. Janet Browne, *Charles Darwin: A Biography*, vol. 1, *Voyaging* (Princeton, NJ: Princeton University Press, 1996); and Browne, *Charles Darwin: A Biography*, vol. 2, *The Power of Place* (Princeton, NJ: Princeton University Press, 2002).

60. Timothy Lenoir, "Essay Review: The Darwin Industry," *Journal of the History of Biology* 20, no. 1 (1987): 115–30.

61. Gould tended to favor professional histories of evolution that agreed with his own interpretations and to dismiss those that did not. In particular, he resisted historical scholarship that argued that progress was an integral theme in Darwin's work. He portrayed Darwin as a break from earlier evolutionary theories, which he saw as progressive. He argued strongly, for instance, that Jean-Baptiste Lamarck represented an earlier, progressive view of evolution. See Gould, *The Structure of Evolutionary Theory*, 170–79. He also argued that Darwin stood apart from his contemporaries, who employed developmental analogies in their understanding of organic transformation. Gould wrote at length on the relationship between growth and evolution in *Ontogeny and Phylogeny*, and he singled out Ernst Haeckel, blaming his comparisons between individual growth and organic transformation for racial and class hierarchies. See Stephen Jay Gould, *Ontogeny and Phylogeny* (Cambridge, MA: Harvard University Press, 1977), 77–79. In *Full House*, Gould dismissed the work of the historian of science Robert Richards, who had argued that development and progress were integral to Darwin's work. Gould dismissed Richards's 1992 book on the subject (which documents the exchange between Darwin and progressive German biology) to bolster Gould's claim that Darwin's class prejudices had led him to imbue natural selection with hints of progress. But Gould wanted to argue that the essentials of the mechanism of natural selection were free from progressive reasoning. Gould, *Full House*, 141; Robert J. Richards, *The Meaning of Evolution: The Morphological Construction and Ideological Reconstruction of Darwin's Theory* (Chicago: University of Chicago Press, 1992).

62. Gould, *Mismeasure of Man*, 321–22.

63. Gould, *The Structure of Evolutionary Theory*, 1342–43.

64. Dennett, *Darwin's Dangerous Idea*, 63.

65. Dennett, 189.

66. Richard Dawkins, *The Extended Phenotype: The Long Reach of the Gene* (Oxford: Oxford University Press, 1999), 19.

67. Dawkins, 35.

68. Dawkins, *Blind Watchmaker*, ix.

69. Dawkins, 5–6.

70. Dawkins, 223–54.

71. Dawkins, 251.

72. Dawkins, 248.

73. Dawkins, 249.

74. Dawkins, 240.

75. Dawkins, 250.

76. Dawkins, 250–51.

77. Stephen Jay Gould to Michael Ruse, 10 April 1989, Box 113, Folder 6, Gould Papers.

78. "Transcript: Debate Between Richard Dawkins and Stephen Jay Gould, Sheldonian Theatre, Oxford University," 1988, Box 897, Folder 17, Gould Papers.

79. The debate between Wilberforce and Huxley took place at the Oxford University Museum of Natural History. Although it was regarded for decades as a classic encounter between Darwinism and religion, historians have painted a more nuanced picture of the exchange. See Frank A. J. L. James, "An 'open clash between science and the church'? Wilberforce, Huxley and Hooker on Darwin at the British Association, Oxford, 1860," in *Science and Beliefs: From Natural Philosophy to Natural Selection, ed.* David M. Knight and Matthew D. Eddy (New York: Routledge, 2005), 171–93; and David N. Livingstone, "Myth 17. That Huxley Defeated Wilberforce in Their Debate over Evolution and Religion," in *Galileo Goes to Jail and Other Myths about Science and Religion, ed.* Ronald L. Numbers (Cambridge, MA: Harvard University Press, 2009), 152–60.

80. Eugenie Scott, director of the NCSE, described this moment as a "state of some flux" because of the "slowing down of creationist public efforts." Eugenie Scott to Stephen Jay Gould, 18 February 1988, Box 93, Folder 17, Gould Papers.

81. Eugenie Scott to Stephen Jay Gould, 14 December 1989, Box 93, Folder 17, Gould Papers.

82. Stephen Jay Gould, "Hooking Leviathan by Its Past," *Natural History* 103, no. 5 (1994): 10.

83. Numbers, *Creationists*, 376–79.

84. Johnson, *Darwin on Trial*, 7.

85. Johnson, 59.

86. Numbers, *Creationists*, 381–82; Barbara Forrest, "The Wedge at Work: How Intelligent Design Creationism is Wedging Its Way into the Cultural and Academic Mainstream," in *Intelligent Design Creationism and Its Critics: Philosophical, Theological, and Scientific Perspectives, ed.* Robert T. Pennock (Cambridge, MA: MIT Press, 2001), 5–50.

87. Center for the Renewal of Science and Culture, "The Wedge," 5 February 1999, http://www.antievolution.org/features/wedge.html.

88. Dennett, *Darwin's Dangerous Idea*, 281.

89. Dennett, 2.

90. Dennett, 50.

91. Dennett, 266.

92. Janet Browne, "Looking at Darwin: Portraits and the Making of an Icon," *Isis* 100, no. 3 (2009): 542–43.

93. Carl Zimmer, *Science Ink: Tattoos of the Science Obsessed* (New York: Sterling, 2011).

94. Jeffrey P. Moran, *American Genesis: The Antievolution Controversies from Scopes to Creation Science* (Oxford: Oxford University Press, 2012), 118–20.

95. Elisabeth A. Lloyd and Stephen J. Gould, "Species Selection on Variability," *Proceedings of the National Academy of Sciences* 90, no. 2 (1993): 595–99; Stephen Jay Gould and Elisabeth A. Lloyd, "Individuality and Adaptation across Levels of Selection: How Shall We Name and Generalize the Unit of Darwinism?," *Proceedings of the National Academy of Sciences* 96, no. 21 (1999): 11904–9.

96. Elisabeth Anne Lloyd, *The Case of the Female Orgasm: Bias in the Science of Evolution* (Cambridge, MA: Harvard University Press, 2005).

97. Stephen Jay Gould, "Freudian Slip," *Natural History* 96, no. 2 (1987): 14–21, reprinted as "Male Nipples and Clitoral Ripples," in Stephen Jay Gould, *Bully for Brontosaurus: Reflections in Natural History* (New York: Norton, 1991), 124–38.

98. Gould, "Freudian Slip," 14.

99. Gould, 16.

100. Gould, 16.

101. John Alcock, "Ardent Adaptationism," *Natural History* 96, no. 4 (April 1987): 4.

102. John Alcock, "Unpunctuated Equilibrium in the *Natural History* Essays of Stephen Jay Gould," *Evolution and Human Behavior* 19 (1998): 321–36.

103. Alcock, 322.

104. Alcock, 333.

105. John Alcock, *The Triumph of Sociobiology* (Oxford: Oxford University Press, 1999).

106. Stephen Jay Gould, "Stephen Jay Gould Replies," *Natural History* 96, no. 9 (April 1986): 4–6.

107. Gould, 4.

108. Gould, 6.

109. Gould, 6.

110. Alcock, "Unpunctuated Equilibrium," 322.

111. Importantly, the denial of Darwinian evolution was not theologically necessary in American Protestantism. James Gilbert argues that evangelicals, particularly at midcentury, had a more complex relationship with science, including evolutionary science, than is generally assumed. Rather than seeing creationism as inevitable, he understands it as a historically emergent phenomenon of the twentieth century. James Gilbert, *Redeeming Culture: American Religion in an Age of Science* (Chicago: University of Chicago Press, 2008). This view is also implied in Numbers's work, particularly in his careful detailing of the various theological engagements with and debates over flood geology. Ronald L. Numbers, *The Creationists: From Scientific Creationism to Intelligent Design*, expanded ed. (Cambridge, MA: Harvard University Press, 2006).

112. Myrna Perez Sheldon and Naomi Oreskes, "The Religious Politics of Scientific Doubt: Evangelical Christians and Environmentalism in the United States," in *The Wiley Blackwell Companion to Religion and Ecology, ed. John Hart* (Hoboken, NJ: John Wiley & Sons, 2017), 348–50.

113. Although evolutionary psychology prompted a popular literature, most feminist engagement with sociobiology and evolutionary psychology was published for professional audiences in fields such as philosophy, the history of science, or evolutionary biology. See, e.g., Marian Lowe, "Sociobiology and Sex Differences," *Signs* 4, no. 1 (1978): 118–25; Linda Marie Fedigan, *Primate Paradigms: Sex Roles and Social Bonds* (Chicago: University of Chicago Press, 1982); Donna J. Haraway, *Primate Visions: Gender, Race, and Nature in the World of Modern Science* (New York: Routledge, 1989); and Diana Tietjens Meyers, "FEAST Cluster on Feminist Critiques of Evolutionary Psychology—Editor's Introduction," *Hypatia* 27, no. 1 (2012): 1–2. By contrast, creation science and intelligent design authors, as well as evolutionary responses, tended to be published by religious or secular trade presses for general audiences. Some examples include Phillip E. Johnson, *An Easy-to-Understand Guide for Defeating Darwinism by Opening Minds* (Downers Grove, IL: InterVarsity, 1997); William A. Dembski, *Mere*

Creation: Science, Faith & Intelligent Design (Downers Grove, IL: InterVarsity, 1998); Kenneth R. Miller, *Finding Darwin's God: A Scientist's Search for Common Ground Between God and Evolution* (New York: HarperCollins, 1999); and Jonathan Wells, *Icons of Evolution: Science or Myth?* (Washington, DC: Regnery, 2000).

114. Andrea Stone and Michael Greshko, "E. O. Wilson, 'Darwin's Natural Heir,' Dies at Age 92," *National Geographic*, 27 December 2021, https://www.nationalgeographic.com/culture/article/eo-wilson-darwins-natural-heir-dies-at-age-92; Carl Zimmer, "E. O. Wilson, a Pioneer of Evolutionary Biology, Dies at 92," *New York Times*, 27 December 2021, Science, https://www.nytimes.com/2021/12/27/science/eo-wilson-dead.html; Scott Neuman, "E. O. Wilson, Famed Entomologist and Pioneer in the Field of Sociobiology, Dies at 92," *NPR*, 28 December 2021, Obituaries, https://www.npr.org/2021/12/27/1068238333/e-o-wilson-dead-sociobiology-entomology-ant-man; Patrick Pester, "Famed Naturalist E. O. Wilson, 'Darwin's Natural Heir,' Dies at 92," livescience.com, 30 December 2021, https://www.livescience.com/e-o-wilson-passes-away; Georgina Ferry, "Edward O. Wilson Obituary," *Guardian*, 6 January 2022, Science, https://www.theguardian.com/science/2022/jan/06/edward-o-wilson-obituary.

115. Monica R. McLemore, "The Complicated Legacy of E. O. Wilson," *Scientific American*, 29 December 2021, https://www.scientificamerican.com/article/the-complicated-legacy-of-e-o-wilson/.

116. Mark Borrello and David Sepkoski, "Ideology as Biology," *New York Review of Books*, 5 February 2022, https://www.nybooks.com/daily/2022/02/05/ideology-as-biology/.

117. Robert H. MacArthur and Edward O. Wilson, *The Theory of Island Biogeography* (Princeton, NJ: Princeton University Press, 1967).

118. Borello and Sepkoski, "Ideology as Biology."

119. E. O. Wilson, "Genes and Racism," *Nature* 289 (1981): 627.

120. Borello and Sepkoski, "Ideology as Biology."

121. Razib Khan, "Setting the Record Straight: Open Letter on E. O. Wilson's Legacy," Substack newsletter, *Razib Khan's Unsupervised Learning* (blog), 18 January 2022, https://razib.substack.com/p/setting-the-record-straight-open.

122. David Sloan Wilson, "A Rebuttal to 'The Complicated Legacy of E. O. Wilson'—This View Of Life," 19 January 2022, https://thisviewoflife.com/a-rebuttal-to-the-complicated-legacy-of-e-o-wilson/.

123. Wilson, *Naturalist*, 15.

Chapter 6 • Is a Democratic Science the End of Western Secularism?

Epigraph: Stephen Jay Gould, *Rocks of Ages: Science and Religion in the Fullness of Life* (New York: Ballantine Pub. Group, 1999), 170.

1. Kristine J. Ajrouch and Amaney Jamal, "Assimilating to a White Identity: The Case of Arab Americans," *International Migration Review* 41, no. 4 (2007): 860–79, https://doi.org/10.1111/j.1747-7379.2007.00103.x; Peter Morey and Amina Yaqin, *Framing Muslims: Stereotyping and Representation after 9/11* (Cambridge, MA: Harvard University Press, 2011), http://www.dawsonera.com/depp/reader/protected/external/AbstractView/S9780674061149; Lila Abu-Lughod, *Do Muslim Women Need Saving?* (Cambridge, MA: Harvard University Press, 2013); Shabana Mir, *Muslim American Women on Campus: Undergraduate Social Life and Identity* (Chapel Hill: University of North Carolina Press, 2014).

2. Sam Harris, *The End of Faith: Religion, Terror, and the Future of Reason* (New York: Norton, 2004) and *Letter to a Christian Nation* (New York: Vintage Books, 2006); Richard Dawkins, *The God Delusion* (London: Bantam, 2006); Daniel. C. Dennett, *Breaking the Spell: Religion as a Natural Phenomenon* (New York: Viking, 2006); Christopher Hitchens, *God Is Not Great: How Religion Poisons Everything* (New York: Twelve Books, 2009).

3. Steven Pinker, *Enlightenment Now: The Case for Reason, Science, Humanism, and Progress* (New York: Penguin Books, 2018), 395.

4. Richard Dawkins, "The Future Looks Bright," *Guardian*, 21 June 2003, https://www.theguardian.com/books/2003/jun/21/society.richarddawkins; Daniel C. Dennett, "Opinion: The Bright Stuff," *New York Times*, 12 July 2003, https://www.nytimes.com/2003/07/12/opinion/the-bright-stuff.html.

5. Talal Asad, *Formations of the Secular: Christianity, Islam, Modernity* (Stanford, CA: Stanford University Press, 2003), 21.

6. On secularism as a mode of governance and the racialization of the American nation, see esp. Tisa Wenger, *We Have a Religion: The 1920s Pueblo Indian Dance Controversy and American Religious Freedom* (Chapel Hill: University of North Carolina Press, 2009); John Lardas Modern, *Secularism in Antebellum America* (Chicago: University of Chicago Press, 2011), 1–48; Kathryn Lofton, *Consuming Religion* (Chicago: University of Chicago Press, 2017), 61–81; Tisa Wenger, *Religious Freedom: The Contested History of an American Ideal* (Chapel Hill: University of North Carolina Press, 2017); and Charles McCray, *Sincerely Held: American Secularism and Its Believers* (Chicago: University of Chicago Press, 2022), 1–28.

7. Gould, *Rocks of Ages*, 5.

8. Gould, 6.

9. Ursula Goodenough, "The Holes in Gould's Semipermeable Membrane between Science and Religion," *American Scientist* 87, no. 3 (1999): 264–68; John Polkinghorne, "The Continuing Interaction of Science and Religion," *Zygon* 40, no. 1 (2005): 43–49.

10. Stephen Jay Gould, "Why I Wrote *Rocks of Ages*," [1999], Box 264, Folder 8, Gould Papers.

11. "The Two Cultures" was published in book form later the same year. See C. P. Snow, *The Two Cultures and the Scientific Revolution* (Cambridge: Cambridge University Press, 1959), 1.

12. Andrew Jewett, *Science, Democracy, and the American University: From the Civil War to the Cold War* (Cambridge: Cambridge University Press, 2012), 302–34.

13. Snow, *Two Cultures*, 2.

14. Steven Shapin and Simon Schaffer, *Leviathan and the Air-Pump: Hobbes, Boyle, and the Experimental Life* (Princeton, NJ : Princeton University Press, 1985); Steven Shapin, "The House of Experiment in Seventeenth-Century England," *Isis* 79, no. 3 (1988): 373–404; Mario Biagioli, *Galileo, Courtier: The Practice of Science in the Culture of Absolutism* (Chicago: University of Chicago Press, 1993).

15. The raced and gendered character of these bodies was highlighted by feminist historians and philosophers of science in the 1980s and 1990s. See, e.g., Londa Schiebinger, *Nature's Body: Gender in the Making of Modern Science* (Boston: Beacon, 1993).

16. Stuart McCook, "'It May Be Truth, but It Is Not Evidence': Paul Du Chaillu and

the Legitimation of Evidence in the Field Sciences," *Osiris*, 2nd ser., 11 (1996): 177–97; Steven Shapin, *The Scientific Life: A Moral History of a Late Modern Vocation* (Chicago: University of Chicago Press, 2008).

17. Henry M. Cowles, *The Scientific Method: An Evolution of Thinking from Darwin to Dewey* (Cambridge, MA: Harvard University Press, 2020), 30–31.

18. Cynthia Eagle Russett, *Sexual Science: The Victorian Construction of Womanhood* (Cambridge, MA: Harvard University Press, 1989), 116–18.

19. Kimberly A. Hamlin, *From Eve to Evolution: Darwin, Science, and Women's Rights in Gilded Age America* (Chicago: University of Chicago Press, 2014), 57–93.

20. Hamlin, 1–24.

21. W. E. B. Du Bois, "The Negro Scientist," *American Scholar* 8, no. 2 (1939): 310.

22. Michel Foucault, *The Birth of the Clinic: An Archaeology of Medical Perception* (New York: Pantheon Books, 1973); Foucault, *The History of Sexuality, Volume I: An Introduction*, trans. Robert Hurley (New York: Random House, 1978); Evelyn C. White, *The Black Women's Health Book: Speaking for Ourselves* (Seattle: Seal, 1990); Sandra Harding, "Is Science Multicultural? Challenges, Resources, Opportunities," *Configurations* 2 (1994): 301–30; Margaret L. Andersen and Patricia Hill Collins, *Race, Class, and Gender: An Anthology* (Belmont, CA: Wadsworth, 1992).

23. Ashis Nandy, ed., *Science, Hegemony and Violence: A Requiem for Modernity* (Delhi: Oxford University Press, 1988); David Arnold, *Colonizing the Body: State Medicine and Epidemic Disease in Nineteenth-Century India* (Berkeley: University of California Press, 1993); Deepak Kumar, *Science and the Raj, 1857–1905* (Delhi: Oxford University Press, 1995); Roy M. MacLeod and Deepak Kumar, eds., *Technology and the Raj: Western Technology and Technical Transfers to India* (Delhi: Sage, 1995); Gyan Prakash, *Another Reason: Science and the Imagination of Modern India* (Princeton, NJ: Princeton University Press,1999); Warwick Anderson, "Introduction: Postcolonial Technoscience," *Social Studies of Science* 32, no. 5/6 (2002): 643–58. For an analysis of the relationship between feminist STS, postcolonial studies, and the practice of science, see the work of evolutionary biologists and the postcolonial feminist theorist Banu Subramaniam in Banu Subramaniam, *Ghost Stories for Darwin: The Science of Variation and the Politics of Diversity* (Urbana: University of Illinois Press, 2014).

24. Paul R. Gross and Norman Levitt, *Higher Superstition: The Academic Left and Its Quarrels with Science*, 2nd ed. (Baltimore: Johns Hopkins University Press, 1998), 4.

25. Norman Levitt, *Prometheus Bedeviled: Science and the Contradictions of Contemporary Culture* (Piscataway, NJ: Rutgers University Press, 1999), 4.

26. Levitt, 5.

27. Levitt, 7.

28. Levitt, 5.

29. Levitt, 7.

30. Ruth Hubbard, "Gender and Genitals: Constructs of Sex and Gender," in "Science Wars," ed. Andrew Ross, special issue, *Social Text*, no. 46/47 (1996): 164.

31. Richard Levins, "Ten Propositions on Science and Antiscience," in Ross, "Science Wars," special issue, 101–11.

32. Sarah Franklin, "Making Transparencies: Seeing through the Science Wars," in Ross, "Science Wars," special issue, 141–55.

33. Richard C. Lewontin, "A la recherche du temps perdu: A Review Essay," in *Science Wars, ed. Andrew Ross* (Durham, NC: Duke University Press, 1996), 293–301.

34. Michael Harris, "Science Wars: The Next Generation," *Science for the People* 22, no. 1 (Summer 2019), https://magazine.scienceforthepeople.org/vol22-1/science-wars-the-next-generation/.

35. Richard Levins and Richard C. Lewontin, *The Dialectical Biologist* (Cambridge, MA: Harvard University Press, 1985). For Hubbard's role in the Genes and Gender Collective, see chapter 2.

36. Alan D. Sokal, "Transgressing the Boundaries: Toward a Transformative Hermeneutics of Quantum Gravity," in "Science Wars," ed. Andrew Ross, special issue, *Social Text*, no. 46/47 (1996): 217–52. Sokal revealed the hoax in Sokal, "A Physicist Experiments with Cultural Studies," *Lingua Franca* 6, no. 4 (1996): 62–64.

37. Bruce Robbins and Andrew Ross, "Mystery Science Theater," *Lingua Franca* 6, no. 5 (1996): 54.

38. Andrew Ross, introduction to "Science Wars," ed. Andrew Ross, special issue, *Social Text*, no. 46/47 (1996): 2–3.

39. Ross, 8.

40. Ross, 4.

41. Stephen Jay Gould, "Deconstructing the 'Science Wars' by Reconstructing an Old Mold," *Science* 287, no. 5451 (2000): 253.

42. Stephen Pope, "Rocks of Ages: Science and Religion in the Fullness of Life," *Christian Century*, 2 June 1999; S. Nomanul Haq, "Thou Shalt Not Mix Religion and Science," *Nature* 400 (1999): 830–31; Goodenough, "Holes in Gould's Semipermeable Membrane."

43. Gould, *Rocks of Ages*, 195.

44. Edward O. Wilson, *Sociobiology: The New Synthesis* (Cambridge, MA: Belknap Press of Harvard University Press, 1975), 547.

45. E. O. Wilson, *Consilience: The Unity of Knowledge*, First Vintage Books Edition (New York: Random House, 1999), 8.

46. Wilson, 201.

47. Wilson, 275.

48. Book jacket to Stephen Jay Gould, *The Hedgehog, the Fox, and the Magister's Pox: Mending the Gap Between Science and the Humanities* (New York: Harmony Books, 2003).

49. Gould, 6.

50. Gould, 6; Gould, *Rocks of Ages*, 4.

51. Gould, *Hedgehog, the Fox, and the Magister's Pox*, 15, 142, 155–56, 236, 258.

52. Stephen Jay Gould, "Nonoverlapping Magisteria," *Natural History* 106 (1997): 16–22, 60–62.

53. Gould, 16.

54. Gould, 16. In his public writing, Gould consistently spoke of creationism as a uniquely American phenomenon. However, twenty-first-century creationism in various forms flourishes many regions outside the United States. Ronald L. Numbers, *The Creationists: From Scientific Creationism to Intelligent Design*, expanded ed. (Cambridge, MA: Harvard University Press, 2006), 399–431.

55. Gould, "Nonoverlapping Magisteria," 16–17.

56. John Paul II, "Truth Cannot Contradict Truth," address to the Pontifical Academy of Sciences, 22 October 1996.

57. Papa Pío XII, *Humani generis* (Valladolid, Spain: Sever-Cuesta, 1950).

58. Gould, "Nonoverlapping Magisteria," 22.

59. Lawrence Badash, "Nuclear Winter: Scientists in the Public Arena," *Physics in Perspective* 3, no. 1 (2001): 76–105; Badash, *A Nuclear Winter's Tale* (Cambridge, MA: MIT Press, 2009); Naomi Oreskes and Erik M. Conway, *Merchants of Doubt: How a Handful of Scientists Obscured the Truth on Issues from Tobacco Smoke to Global Warming* (New York: Bloomsbury, 2010), 47–54; Myrna Perez Sheldon, "Stephen Jay Gould and the Value of Neutrality of Science During the Cold War," *Endeavour, Science in the Public Eye* 40, no. 4 (2016): 248–55.

60. Gould, "Nonoverlapping Magisteria," 61.

61. Chris Lavers, "A Hands-off God?," *Guardian*, 2 February 2001; Mark W. Durm and Massimo Pigliucci, "Rocks of Ages: Science and Religion in the Fullness of Life," *Skeptical Inquirer*, November 1999, 53.Durm and Pigliucci's review presented opposing viewpoints by the two authors; the quotation is from Pigliucci.

62. Richard Dawkins, quoted in Steve Paulson, *Atoms and Eden: Conversations on Religion and Science* (Oxford: Oxford University Press, 2010), 10.

63. E. O. Wilson, quoted in Paulson, 10.

64. Gould, "Why I Wrote *Rocks of Ages*."

65. Wendy L. Wall, *Inventing the "American Way": The Politics of Consensus from the New Deal to the Civil Rights Movement* (Oxford: Oxford University Press, 2008), 8–10.

66. Gould, "Why I Wrote *Rocks of Ages*."

67. Gould, *Rocks of Ages*, 4.

68. A notable exception is Richard York and Brett Clark, *The Science and Humanism of Stephen Jay Gould* (New York: Monthly Review Press, 2011).

69. Harris, *End of Faith*, 109.

70. Dawkins, *God Delusion*, 57.

71. Harris, *End of Faith*, 131.

72. Harris, 107.

73. Hitchens, *God Is Not Great*, 136–37.

74. Harris, *End of Faith*, 132.

75. Dennett, *Breaking the Spell*, 13.

76. Jasbir K. Puar, *Terrorist Assemblages: Homonationalism in Queer Times* (Durham, NC: Duke University Press, 2007), xii.

77. Amaney A. Jamal and Nadine Christine Naber, eds., *Race and Arab Americans before and after 9/11: From Invisible Citizens to Visible Subjects (Syracuse, NY: Syracuse University Press, 2008)*, 1–2.

78. Minoo Moallem, "Whose Fundamentalism?," *Meridians* 2, no. 2 (2002): 298.

79. Stephen Jay Gould, *I Have Landed: The End of a Beginning in Natural History* (New York: Harmony Books, 2002), 387.

80. Stephen Jay Gould, "A Time of Gifts," *New York Times*, 26 September 2001, A19.

81. Gould, *I Have Landed*, 387.

82. Stephen Jay Gould, *Triumph and Tragedy in Mudville: A Lifelong Passion for Baseball* (London: Cape, 2004), 37.

83. Gould, *Rocks of Ages*, 7.

84. Sandra G. Harding, *The Feminist Standpoint Theory Reader: Intellectual and Political Controversies* (New York: Routledge, 2003).

85. Gould, *Rocks of Ages*, 4.

86. Gould, *I Have Landed*, 401.

87. Gary Wolf, "The Church of the Non-Believers," *Wired*, 1 November 2006, https://www.wired.com/2006/11/atheism/.

88. Matthew Avery Sutton, *American Apocalypse: A History of Modern Evangelicalism* (Cambridge, MA: Belknap Press of Harvard University Press, 2014), xiv.

89. Linda Villarosa, "Why America's Black Mothers and Babies Are in a Life-or-Death Crisis," *New York Times Magazine*, 11 April 2018, https://www.nytimes.com/2018/04/11/magazine/black-mothers-babies-death-maternal-mortality.html; Aimee Cunningham, "What we can learn from how a doctor's race can affect Black Newborns' Survival," *ScienceNews*, 25 August 2020, https://www.sciencenews.org/article/black-newborn-baby-survival-doctor-race-mortality-rate-disparity.

90. Twitter thread #blackbirders, accessed April 2021, https://twitter.com/hashtag/blackbirders?lang=en.

91. Robbie Wheelan, "Native-American Tribes Pull Ahead in Covid-19 Vaccinations," *Wall Street Journal*, 10 April 2021, https://www.wsj.com/articles/native-american-tribes-pull-ahead-in-covid-19-vaccinations-11618047001.

92. On the "close collaboration" between a program at Simon Fraser University and the Tla'amin Nation, see Clarissa Yap, "Pi Day: How Indigenous Stone-Fish-Traps Contribute to Mathematical Education," *Simon Fraser University News*, 12 March 2021, https://www.sfu.ca/sfunews/stories/2021/03/pi-day--how-indigenous-stone-fish-traps-contribute-to-mathematic.html.

Epilogue • Why We Can't Wait for a Democratic Science

1. Catherine Schoichet, "Covid-19 Is Taking a Devastating Toll on Filipino American Nurses," *CNN*, 11 December 2020, https://www.cnn.com/2020/11/24/health/filipino-nurse-deaths/index.html.

2. Catherine Ceniza Choy, *Empire of Care: Nursing and Migration in Filipino American History* (Durham, NC: Duke University Press, 2003), 1–14.

3. A memorial site was created to memorialize Filipino health care workers who have died during the global COVID-19 pandemic. It is called Kanlungan, which means "shelter" or "refuge." https://www.kanlungan.net/.

4. Julie McCarthy, "A Filipina Nurse on Working on the Front Lines of the Pandemic," *All Things Considered, NPR*, 7 September 2020.

5. Jessie Yeung and Xyza Cruz Bacani, "When Love Is Not Enough: Millions of Filipinos Work Abroad to Support Their Families Back Home. Eight Mothers and Children ;Share Their Stories of Separation," *CNN*, November 2020, https://www.cnn.com/interactive/2020/11/asia/hong-kong-filipino-helpers-dst/.

6. Stephen Jay Gould, *I Have Landed: The End of a Beginning in Natural History* (New York: Harmony Books, 2002), 21–25; Gould, *Rocks of Ages: Science and Religion in the Fullness of Life* (New York: Ballantine Books, 1999), 7–8; Gould, *The Mismeasure of Man*, rev. and expanded ed. (New York: Norton, 1996), 38–39. Gould dedicated *The Mismeasure of Man* to "the memory of Grammy and Papa Joe."

7. He knew that both sets of his grandparents were Eastern European Jewish

immigrants to New York City, and he knew the details about his maternal grandmother's immigration from Hungary through Austria to work as a seamstress in a "sweatshop in the lower east side." Stephen Jay Gould, "For Jesse and Ethan from Daddy," 26 December 1976, Box 667, Folder 8, Gould Papers.

8. See, e.g., Audre Lorde, *Sister Outsider: Essays and Speeches* (Berkeley, CA: Crossing, 1984); and Ta-Nehsi Coates, *Between the World and Me* (New York: Spiegel & Grau, 2015), 107.

9. Martin Luther King Jr., *Why We Can't Wait* (New York: New American Library, 1964).

10. Walter Johnson, *River of Dark Dreams: Slavery and Empire in the Cotton Kingdom* (Cambridge, MA: Harvard University Press, 2013), 1–45.

11. Tisa Wenger, *Religious Freedom: The Contested History of an American Ideal* (Chapel Hill: University of North Carolina Press, 2017), 15–53.

12. Spencer Tucker, *The Encyclopedia of the Spanish-American and Philippine-American Wars: A Political, Social, and Military History, 3 vols.* (Santa Barbara, CA: ABC-CLIO, 2009), 1:504–5.

13. Theodore Roosevelt, *The Strenuous Life: Essays and Addresses* (New York: Century, 1901), 11. "The Strenuous Life" was a speech given by Roosevelt in Chicago in 1899.

14. Paul A. Kramer, *The Blood of Government: Race, Empire, the United States, and the Philippines* (Chapel Hill: University of North Carolina Press, 2006).

15. Odd Arne Westad, *The Global Cold War: Third World Interventions and the Making of Our Times* (Cambridge: Cambridge University Press, 2007), 1–38.

16. Laura Wexler, *Tender Violence: Domestic Visions in an Age of U.S. Imperialism* (Chapel Hill: University of North Carolina Press, 2000); Julian Go, *American Empire and the Politics of Meaning: Elite Political Cultures in the Philippines and Puerto Rico during U.S. Colonialism* (Durham, NC: Duke University Press, 2008); David Brody, *Visualizing American Empire: Orientalism and Imperialism in the Philippines* (Chicago: University of Chicago Press, 2010); Karine V. Walther, *Sacred Interests: The United States and the Islamic World, 1821–1921* (Chapel Hill: University of North Carolina Press, 2015); Jasbir K. Puar, *Terrorist Assemblages: Homonationalism in Queer Times* (Durham, NC: Duke University Press, 2017).

17. Combahee River Collective, "Combahee River Collective Statement," in *How We Get Free: Black Feminism and the Combahee River Collective, ed.* Keeanga-Yamahtta Taylor (Chicago: Haymarket Books, 2017), 16.

18. Sara Moslener, *Virgin Nation: Sexual Purity and American Adolescence* (New York: Oxford University Press, 2015), 130–54.

19. Wayne Grudem and John Piper, eds., *Recovering Biblical Manhood and Womanhood: A Response to Evangelical Feminism* (Wheaton, IL: Crossway Books, 1991); Myrna Perez Sheldon, "Wild at Heart: How Sociobiology and Evolutionary Psychology Helped Influence the Construction of Heterosexual Masculinity in American Evangelicalism," *Signs: Journal of Women in Culture and Society* 42, no. 4 (2017): 977–98.

20. Tim LaHaye and Jerry B. Jenkins, *Left Behind: A Novel of the Earth's Last Days*, Left Behind Series (Wheaton, IL: Tyndale House, 1995); Joshua Harris, *I Kissed Dating Goodbye* (Sisters, OR: Multnomah Books, 2003); Eric Ludy and Leslie Ludy, *When God Writes Your Love Story: The Ultimate Approach to Guy/Girl Relationships* (New York: Doubleday Religious Publishing, 2007).

INDEX

abortion, 10, 38, 51, 55–56, 77, 79, 109, 110–14, 128, 190
academic Left, 18, 49, 54, 77, 154, 176, 177, 179; alignment with creationism, 128, 136, 196n43; democratizing science goal, 181–82; science criticism, 19, 128, 166, 175, 181–82; sociobiology criticism, 54, 80, 154; support for scientific liberalism, 80. *See also* leftist perspectives, on science
Acts and Facts, 114, 127, 189–90
adaptationism, Gould's criticism, 13, 14, 89–97, 104, 191, 214n33; adaptive sexuality, 97–98, 151–54; criticisms of Gould's position, 14–15, 145–54; Panglossian paradigm critique, 13, 61, 81–84, 90–91, 92, 97, 129, 134. *See also* punctuated equilibria model
Adapted Mind, The, 98–99
Adler, Jerry, 139
Agassiz, Louis, 1
Alcock, John, 152–53, 154; *The Triumph of Sociobiology,* 152–53
Alexander, Michelle, *The New Jim Crow . . .,* 45–46, 48
Allen, Garland, 63–65
altruism, 86–87, 100; reciprocal, 61
Alvarez, Luis Walter, 94
American Academy for the Advancement of Science, 27, 38
American Anthropologist, 31
American Anthropology Society, 31
American Association for the Advancement of Science, 63, 220n42
American Civil Liberties Union, 107–8, 115, 124
American democracy: Gould's perspective, 7, 17, 173–74, 178–80, 190–91; postcolonial narrative, 188–91
American democracy, relationship to science, 7–11, 15–16, 132; Black perspective, 37–41; during Cold War, 8, 129, 130–31 ; Gould's perspective, 3, 52, 161–62, 173–74, 179, 180; implication of scientific racism for, 7, 25–30; leftist perspective, 24–25, 50–51, 52, 131, 132; New Atheism's perspective, 159–60, 176; right-wing perspective, 131; scientific authority debates, 109, 130, 131; scientific liberalism's perspective, 49, 50–52, 130, 131–32, 165–67; scientific racism and, 25–30; secularism and, 159–62, 192
American Museum of Natural History, 34, 42, 51, 56
American Reformed Baptists, 102–3
American Scholar, 103–4
American School of Ethnology, 1, 63
anthropology/anthropologists, 1, 2, 4, 31, 56, 57, 60, 85–86
Antioch College, 22, 31–32
Ardrey, Robert, 50; *African Genesis,* 25
Asad, Talal, 160
Association for Women in Science, 57
authority, scientific, 10–11, 19, 135–36; for biological determinism, 73–74, 76; Christian Right's challenges, 135–36; creationist challenges, 15, 16–17, 106, 109, 118, 122, 127–28, 160–61, 187; critical humanist perspective, 163, 174–76; Darwin's influence on, 140–41; of democratized science, 187, 191–92; of evolutionary science, 55, 127,

authority, scientific (*cont.*)
140–41; evolution-creation debate and, 139–40; feminist criticism, 73–74, 75, 101, 160, 161, 174, 187; Gould's perspective, 68, 161–62, 174, 186, 191; leftist critiques and disruptions, 10–11, 15, 27, 38–39, 53, 59, 79, 128, 135–36, 160–61, 163; as moral authority, 41, 170–71, 173, 174; public skepticism about, 168; relation to American democracy, 109, 130, 131; scientific liberalism's defense, 3, 10, 15, 51–52, 68, 101–2, 130; scientific populism and, 126–32, 187; in scientific racism debates, 26, 36–37, 68; of sociobiology, 52, 53, 57, 80, 154–58, 187; whiteness-based, 80, 164
Ayala, Francisco, 116–17, 124, 220n42

Balanced Treatment for Creation-Science and Evolution Science Act, 107–8, 115
Barber, Edwin, 43, 64
Beauvoir, Simone de, *The Second Sex*, 74
Beckwith, Jonathan, 53
behavioral ecology, 87
Behe, Michale, *Darwin's Black Box*, 133
Bell, Derrick, 39–40
Benjamin, Ruha, *Race After Technology . . .*, 48–49
Berlin, Isaiah, 171
"biogenetic law," 66
biological determinism, 27, 52–53, 64; academic left's criticism, 181–82; as conservative political ideology, 5, 28, 58, 76; feminist criticism, 5, 38, 70–71, 78; Gould's criticism, 5, 35–36, 60–65, 78; limits of criticism of, 74–80. *See also* sociobiology
biopolitics, 76–78, 128–31, 188, 189
Black Americans: achievement disparities, 26; lack of political power, 24; poverty, 25–26; scientific exploitation, 1, 4–5, 6, 7, 8, 78, 193n1; as scientists, 103–4, 165; self-help activism, 47; social status, 77, 155–56. *See also* civil rights movement
Black critiques, of science, 37–41
Black Lives Matter movement, 46
Black Panther Party, 8, 23–24, 28, 38–39, 47, 79, 128, 131, 182
Black Power politics, 35, 37–41, 47
Bogue, Grant, 36
Borello, Mark and David Sepkoksi, "Ideology as Biology," 155, 156, 157
Boston College, 58
Boston Women's Health Book Collective, 128
Briscoe, Anne, 57
Bryan, William Jennings, 119, 127
Bulletin of the Atomic Scientists, 27–30
Burgess Shale, 95–96, 105
Burnet, Thomas, 95
Bush, George W., 159
Butler, Judith, *Gender Trouble . . .*, 73–74

Cambridge Forum, 123
Cambridge University, 162
Carter, Jimmy, 44, 96, 105–6, 111, 115–16
Catholicism, 81, 113, 161, 171–73, 176, 188
Center for the Renewal of Science and Culture, 148
Central State College, 22, 34
Chaney, James, 34–35
Christian evangelicalism, 2–3, 4, 10, 84, 109–11; apocalyptic vision, 181, 190; complementarianism movement, 102–3, 190; ethnonationalism focus, 7, 11, 46, 79–80, 114, 131, 157, 189, 190. *See also* Christian Right
Christian fundamentalism, 19, 106, 117, 172, 218n8
Christian Heritage College, 113, 114, 189–90
Christian Right, 2–3, 110–11, 112, 122–24, 189; antiabortion activism, 109, 110–14; creationism and, 9–10, 108–16, 126–27, 154, 229n111; culture wars, 103, 196n42; political identity and power, 108–9, 123, 126–27, 154, 218n8; populism, 126–32
Christianson, Eric, 117
civilization, evolutionary and developmental models, 130, 175–76, 205n16
Civil Rights Act (1964), 23
civil rights movement, 2, 16, 24, 25, 46, 75, 201n64; direct-action tactics, 32–33, 35, 47, 54; Gould's participation, 2, 22–23, 24, 32–33, 34, 35, 37, 47, 50, 54, 198n2
climate change denialism, 7, 8, 187
Cold War, 8–9, 77, 128, 129–32, 169, 188, 195n30
colonialism/colonization, 4–5, 13–14, 17, 38, 77, 129, 160, 182–83, 187–89, 191

Columbia University, 22, 31, 33, 57
Combahee River Collective, 37–38, 39, 47, 75, 78, 188
Committee Against Racism, 63, 208n68
Committees of Correspondence, 124–25
communism, 8, 30, 129–30, 188
Concerned Women for America, 114
Congress on Racial Equality (CORE), 32–33, 34–35, 47
conservatives/conservatism, 36, 40–41, 51, 58, 108–9
contingency, historical, 14, 18, 83, 84, 96–97, 103–6, 217n99
contraception, 6–7, 51, 55–56, 76–77, 123
Council for Democratic and Secular Humanism, 140
Council on Biblical Manhood and Womanhood, 102–3, 190; *Recovering Biblical Manhood and Womanhood*, 103
counterculture, 8, 195n29
counterfactuals, in history, 103–4
COVID-19 pandemic, 7, 185–86; vaccines, 7, 8, 185
craniometry, 1–2, 3–5, 7, 64, 65, 74, 165, 193n1
creationism, 2–3, 105–6, 107–32, 128; as American phenomenon, 172, 233n54; Christian Right and, 9–10, 108–16, 126–27, 154, 229n111; as political movement, 115–16, 121, 134; punctuated equilibria model and, 134–40
creationism-evolution debate, 2–3, 15, 128, 147–48, 150, 153, 181; evolutionary biologists' participation, 18, 115–26, 132, 133, 161, 163, 176; Gould's participation, 16, 108, 117–26, 134, 169, 172; public's exposure to, 153–54; punctuated equilibria model and, 18, 96, 136–40; Scopes (Monkey) trial, 107–10, 118–19, 120–21, 122, 127, 218n6; young-earth creationists and, 9–10, 113–14
creation science, 9–10, 113–14, 131, 196n36; populism of, 126–32
creation science, in school curricula, 115–16, 124–25, 154, 187; *Edwards* case, 108, 147; *Kitzmiller* case, 133–34; *McLean* case, 18, 107–9, 114–21, 122–26, 132, 136, 139, 140, 145, 147, 150, 176, 201n62
Crenshaw, Kimberlé, 78–79, 185
critical race theory, 39–40, 41, 157–58, 160, 161, 163, 191
culture wars, 18–19, 103, 196n42

Daly, Martin and Margo Wilson, *Sex, Evolution, and Behavior*, 103
Darrow, Clarence, "Evolution, A 'Crime,'" 120–21
Darwin, Charles, 18–19, 30, 149; centennial celebrations, 140–41, 143; Darwin studies, 143, 149, 226n59; depictions in popular culture, 140, 150; New Atheism's perspective on, 159–60; *On the Origin of Species*, 141, 150, 159–60; Victorian cultural context, 59, 69, 140, 142–43, 227n61
Darwinian evolutionary theory, 2–3, 11, 18–19, 93, 133, 141, 142–43, 225n41; algorithmic nature of, 149–50, 153; connection with creationism, 144; creation science's criticism, 114–15; Dawkins-Gould debate on, 144–47; Gould's perspective, 10, 83–84, 97, 134, 136–37, 139, 142–50, 180, 227n61; gradualism, 136–37, 146. *See also* adaptationism; natural selection
"Darwin Wars," 139–40
Dawkins, Richard, 139, 144–45, 159, 160, 173, 175, 180, 181; *The Blind Watchmaker*, 14, 127, 134–35, 145, 146, 153, 176; *The Extended Phenotype*, 145; *The God Delusion*, 175, 176; Oxford University debate with Gould, 146–47; *The Selfish Gene*, 9, 12, 51, 55, 56, 86, 88, 101, 105, 145, 151, 176
Democratic Party, 31, 46
democratic science, 15, 17, 41–42, 187; cosmological unity and, 181–84; need for, 185–92
Dennett, Daniel, 139, 148, 159, 160, 180; *Breaking the Spell*, 177; *Darwin's Dangerous Idea . . .*, 135, 144, 149–50, 153
developmental biology, 53, 66, 67, 83, 91, 142, 151–52, 153, 208n83, 227n61
Dignity and Power Now, 48
dinosaurs, extinction, 83, 94–95
Discover, 122, 127, 138–39, 140; "Darwin on Trial . . . ," 140
Discovery Institute, 133, 148
DNA, 85, 87
Dobson, James, 102

Dobzhansky, Theodosius, 36, 141
Du Bois, W. E. B., 32–33, 77, 165; "The Negro Scientist," 103–4
Duke University, 167
Durant, John, 146–47

Eldredge, Niles, 93–94, 134–35, 137–38, 147
empiricism, 8, 28, 40, 132, 155, 164, 181, 192; in creation science, 127; in evolutionary science, 15–16, 67, 86, 89–91, 109, 122, 127, 132, 137, 144, 152, 153; feminist science perspective, 12, 89; in genetics, 141; Gould's perspective, 67, 68, 75, 89–90, 92, 122, 137, 144, 153, 161, 172; of New Atheists, 181; in paleobiology, 94, 153; religious authority *versus*, 109, 120, 127, 161, 172, 173; scientific racism and, 1, 4, 28, 48, 68; in sociobiology, 50–51, 57, 59, 61, 80, 86, 87–88, 89–90, 92, 94; trustworthiness in, 164
Enlightenment science, 7, 10, 11, 15, 80, 132, 158, 159–60, 164, 166, 170, 181–82, 183, 187
environmentalism, 8, 77, 123, 128–29, 136, 155, 189, 195n30
epistemology, scientific, 11, 14, 20, 38, 59, 78, 90, 165; denaturalization, 181–82; feminist, 59, 60, 68, 74, 76; populist, 17; postmodern, 167–68; sexism in, 56, 60
Equal Rights Amendment, 51, 102, 109, 114, 121
ethical science, 3, 20, 112–13, 155, 163, 170, 181, 186; Gould's perspective, 9, 62, 63, 66, 84, 90, 170, 171
ethnonationalism, 7, 11, 19, 46, 79–80, 114, 131, 157, 189, 190
Ethology and Sociobiology, 156
eugenics, 7, 25, 53, 64, 65, 77–78, 113, 132
Evolution, 141
Evolution and Human Behavior, 152
evolutionary biology, 3, 11–12, 16; empiricism, 15–16, 67, 86, 89–91, 109, 122, 127, 132, 137, 144, 152, 153; E. O. Wilson's perspective, 62, 155, 170–71; feminist criticism, 3, 5; institutionalization, 117. *See also* adaptationism; creationism-evolution debate; Darwinian evolutionary theory; natural selection
Evolutionary Biology, 37
evolutionary synthesis, 16, 140–42, 146, 149, 154, 215n64, 225n41
evolutionary theory: gene-centered, 86–87, 112, 145; Gould's revolutionization of, 34; hierarchical/progressive view, 13, 66–67, 83–84, 95–96, 97, 144, 197n58, 217n104, 227n61; pluralistic, 151. *See also* Darwinian evolutionary theory
Evolution & Development, 4
expertise, scientific, 7, 8, 9, 28–29, 79–80, 108–9, 117, 125, 167; scientific populism *vs.*, 126–32
extinctions, 13, 83, 92, 94–95, 96, 105

Falwell, Jerry, 111, 113, 119, 198n64; *Listen America!* sermons, 113
Farmer, James, 32–33, 35, 200n47
Fasteau, Brenda Feigen, 69
femininity, 84
feminism, 8, 15, 20, 75, 103, 179, 190; Black, 37–38, 39, 48–49, 74–76; Christian evangelicals' opposition, 102–3, 190; liberal criticism, 135; racism within, 38; socialist, 38; sociobiological perspective, 99; white, 74–75
feminist science, 2, 7, 29, 52, 68–74, 157–58, 170, 195n33; criticisms of biological determinism/sociobiology, 5, 9, 51, 56–60, 74, 135, 150, 187; criticisms of science, 49, 70–71, 101, 160, 161, 165; epistemology, 59, 60, 68, 74, 76; evolutionary psychology's criticisms of, 101; science wars and, 167; scientific liberalism's criticisms of, 74, 157, 176–77, 196n43
Field Museum of Natural History, 137–38, 224n20
Filipino Americans, 20, 185–88
First Amendment, establishment clause, 108, 115, 117
Focus on the Family, 102
fossil record, 92–94, 95–96, 105, 148, 153. *See also* punctuated equilibria model
Foucault, Michel, *History of Sexuality*, 76
Fox, Sidney, 124
Franklin, Sarah, 167
Frazier, Demita, 37–38
Frazier, Nishani, *Harambee City . . .*, 47
Friedan, Betty, *The Feminine Mystique*, 114

Garlow, Jim, 190
gay rights, 10, 110–11, 113, 114, 178, 190

Gegner, Lewis, 22–23, 33, 34–35
gender: biological binary view, 84, 167; feminist perspective, 56–57, 72–74, 167; identity, 41, 72; sociobiological perspective, 84; studies, 178
gender roles, 130–31; biological determinism theories, 5, 9, 54, 56–57; Christian evangelicals' perspective, 102–3, 190; conservatives' perspective, 51, 89; evolutionary theory, 9, 54, 99–100; in families, 131; heteronormative, 108; of homosexuals, 99–100; sociobiological theory, 12, 53–54, 55, 57, 82, 85, 97, 176; traditional, 51, 89. *See also* sexism, scientific; sociobiology
Genes and Gender Collective, 18, 56–59, 70, 75, 78, 99, 167; *Genes and Gender*, 56–59, 70–71, 76
genetics: as challenge to Darwinism, 141; Mendelian, 87, 141, 225n42; molecular, 27–28, 83, 87, 145; population, 37, 86–87, 141, 225n42, 225n.42, 226n.42; of speciation, 142
genocide, 5, 182–83, 189, 191, 192
geology: flood geology, 113–14, 127, 175, 229n111; historical, 95
Gish, Duane, 10
Gliddon, George, 1
Godfrey, Laurie R., 124
Goodman, Andrew, 34–35
Gorman, James, "The Tortoise and the Hare," 138–39
Gould, Deborah, 107, 119
Gould, Ethan, 107
Gould, Jesse, 107
Gould, Leonard, 30–31
Gould, Stephen Jay: *Biological Determinism* (unpublished work), 63–65; biological universalism theory, 36–37, 39–40; "Creationism: Genesis *vs.* Geology," 123; "Deconstructing the 'Science Wars.'..," 169; *Eight Little Piggies*, 203n92; *Ever Since Darwin*, 43–44, 138, 203n92; "Evolution as Fact and Theory," 122, 133; "Evolution Wins Again," 120–21; final illness and death, 159, 221n68; *Full House* . . . , 97; geology/paleontology background, 66; *The Hedgehog, the Fox, and Magister's Pox* . . . , 170, 171, 174; Jewish immigrant heritage, 14, 30, 179, 186–87, 190–91, 200n33, 235n7; McLean trial testimony, 107, 118–21, 118–22, 125–26, 201n62; *Models in Paleobiology*, 93; "Moon, Mann, and Otto," 118–19; "Mysteries of Evolution," 139; NCSE and, 125–26; *Ontogeny and Phylogeny*, 66, 67, 69, 89, 90–91, 138, 208n83, 227n61; *A Panda's Thumb*, 44; pluralistic approach, 152, 154, 161; public recognition, 42–45, 96–97, 139; "Racist Arguments and I.Q.," 22; relationship with E.O. Wilson, 61–63, 64, 65; *Rocks of Ages* . . . , 159, 161–62, 169–70, 171, 174; SSG membership, 53, 63; *Structure of Evolutionary Theory*, 44–45, 138; "The Evolutionary Biology of Constraint," 81; "The Nonscience of Human Nature," 50; "The Paradox of the First Tier . . . ," 94–95; "The Spandrels of San Marco and the Panglossian Paradigm" (Lewontin, co-author), 61, 81–84, 90–91, 92, 97, 134, 149, 197n57; *Time's Arrow, Time's Cycle* . . . , 95, 215n56; "Why I Wrote *The Rock of Ages*," 173–74; *Wonderful Life*. . . , 13–14, 44, 95–96, 97, 104–6, 215n62
Grabiner, Judith V., 119
Gross, Paul R. and Norman Levitt: *Higher Superstition* . . . , 166, 167, 168; *Prometheus Bedeviled* . . . , 166

Hampton, Fred, 28
Haraway, Donna, 70, 73
Harding, Sandra, 73
Harris, Samuel, 159, 175, 180, 181; *The End of Faith*, 175–76
Harvard Educational Review, 26
Harvard Review, 28, 29
Harvard University, 9, 50, 53, 63–64, 67, 85, 147; Department of Genetics, 152; Gould's professorship, 2, 30, 33, 36, 37, 42, 44–45, 96–97, 141–42, 179; Law School, 39
Head Start, 26
health care and medicine: leftist collectives, 6, 8, 23–24, 37, 38, 55–56, 59–60, 128, 131, 160, 182, 191; male domination, 55–56, 60, 165; professionalization, 4; racial diversity, 182; racial segregation/racism in, 8–9, 23–24, 37, 38; racist practices, 4–5, 38, 182, 183, 185; sexist practices, 55, 165; women's health movement, 37, 49, 55–56, 59–60, 89, 99, 128

Heritage Foundation, 114
Herlihy, Mark, 120
heterosexuality, 38, 73, 154, 187; conservative Christian perspective, 79–80, 102–3, 154, 187, 190; evolutionary psychology's perspective, 99–100; feminist perspective, 38, 73
historical literacy/reflection, 69, 70, 74
historically Black colleges and universities, 22
historicity/historical contingency, 14, 18, 83, 84, 92–93, 94–96, 103–6
history of science, 65; feminist research, 69–70, 73–74, 229n115; Gould's interest in, 34, 42–43, 44, 63–64, 69; history of biology, 64
Hitchins, Christopher, 159
homonationalism, 177–78
homophobia, 75, 102
homosexuality, 99–100, 113
hormones, 6–7, 57, 64, 70
Hubbard, Ruth, 2, 5, 51, 57–58, 80, 167; *Genes and Gender* co-editorship, 59; "Have Only Men Evolved?" 59; "New Reproductive Technologies," 59–60; *The Politics of Women's Biology*, 58
humanism, secular, 113, 140–41
humanistic approach, to science, 5, 20, 28, 30–31, 68, 159–60, 170, 171, 182; Christian Right's criticism of, 112–13, 114; of feminist science, 29, 56, 58, 72–73; Gould's perspective, 18, 68, 170–75, 202n84; as leftist criticism, 28–29; science wars and, 19, 76, 162–69, 174–75, 176–77
hunter-gatherer societies, 12, 53–54, 57, 98, 99, 101
Hutton, James, 95
Huxley, Thomas Henry, 136–37, 147, 228n79
Hypatia, 71–72, 73–74
hypersexuality, 100, 101

identity: politics of, 19; role in science, 164–65; scholarly, 178–79; of scientists, 23; Western, 154, 157
immigrants/immigration, 15, 20, 45, 46, 77, 132, 174; demographic changes and, 189–90; Gould's immigrant heritage, 14, 30, 179, 186–87, 190–91, 200n33, 235n7; immigration regulation, 77–78, 189; political power, 24; rights, 114
incarceration/incarcerated populations, 20, 45–46, 48, 189, 203n102
Indiana University, 72–73
indigenous perspective, on science, 157–58, 165, 183
Institute for Creation Research, 113–15, 127, 147, 189–90
Institute of Women's History, 69–70
intelligence quotient (IQ)-race debate, 25–30, 35–36; craniometry studies, 1–2, 3–5, 64, 65, 67–68; Gould's perspective, 65–68
intelligent design, 15, 133–34, 135, 139, 145, 147, 148, 149–50, 153
interdisciplinary scholarship, 162–63, 165, 174–75
Islam, 15, 159, 160, 175–76, 177, 191

Jaroff, Leon, 138
Jensen, Arthur, 76; "How Much Can We Boost IQ . . . ," 26, 35; Lewontin's criticism of, 27–30
Johnson, Lyndon B., 40–41
Johnson, Phillip, *Darwin on Trial*, 133, 147–48

Kakutani, Michiko, 7–8
Kansas State School Board, 133–34
Keel, Terence, 48
Keller, Evelyn Fox, 71, 73; *A Feeling for the Organism . . .* , 72
Kennedy, John F., 130
Kennedy, John F., Jr., 123
Kennedy, Robert, 33
Khan, Razib, 157
King, Martin Luther, Jr., 33, 35, 37, 46–47, 113, 201n64
kinship practices, 48–49, 57, 87, 99
Kohlstedt, Sally Gregory, 70, 71
Krushchev, Nikita, 130–31
Kuhn, Thomas, 67, 89–90, 143–44
Kurtz, Paul, 140–41

Lacks, Henrietta, 6, 7
LaHaye, Beverly, 114
LaHaye, Timothy, 113–14, 127, 189–90, 198n64; *Left Behind* series, 190
Latour, Bruno, 10–11, 19
Leacock, Eleanor, 57, 59
Leakey, Richard E., 115; *Origins*, 43

leftist critique, of science, 3, 5, 8–9, 10, 15–16, 106, 131, 175, 183; creationism and, 109, 132, 135–36, 160–61; democratization of science, 182–83; feminist, 49, 52, 56–57; generative nature of, 157–58; Gould's perspective, 16, 17, 18, 24, 30, 31–32, 36–37, 40, 42, 50, 75; public shaming tactics, 41, 53–55, 80, 156; relation with scientific liberalism, 162, 165–67, 177; role in American democracy, 17; science wars, 19, 162–69, 174–75, 176–77, 179; scientific authority and, 10–11, 15, 27, 38–39, 53, 59, 79, 128, 135–36, 154–55, 160–61, 163; tactics, 28–29. *See also* leftist criticism *under specific topics;* radical science movement

leftist movements and politics: Black, 8, 23–24, 28, 47, 79, 128, 131, 182; health collectives, 8, 23–24, 38, 79, 128, 131, 160, 182, 191. *See also* New Left

Levins, Richard, 28, 167

Levitt, Norman, *Prometheus Bedeviled . . .*, 166–67, 168

Lewin, Roger, 115, 117–18, 124, 167; "A Response to Creationism Evolves," 115, 124–25

Lewis, Jason, 2, 4

Lewontin, Richard, 37, 39–40, 54, 59, 91; critique of Arthur Jensen, 27–30, 35, 151; "Race and Intelligence," 28; "The Apportionment of Human Diversity," 37; "The Spandrels of San Marco" (co-author with Gould), 61, 81–84, 90–91, 92, 97, 134, 149, 197n57

LGBTQ+ community, 5, 78, 178; Black lesbian feminists, 37–38, 39, 74. *See also* queer studies/theory

liberal approaches, to science, 131–32. *See also* liberalism, scientific

liberalism, 122–23, 130, 177–78, 188–89; color-blind, 35, 37, 40–41; ethnonationalism narratives, 177–78, 188–89; religious, 175; support for evolutionary science, 154

liberalism, scientific, 11–12, 131, 182; apoliticism, 163; creationism and, 15, 109, 135; as defense against ethnonationalism, 80; democratization of science role, 182; of E. O. Wilson, 9, 54–55, 60–63, 80, 109, 165–66; epistemic values, 29; evolutionary science and, 3, 154; Gould's position, 16, 68; interdisciplinary challenges to, 165; race/scientific racism perspectives, 26–28, 35, 37, 40–41, 68, 165; radical science movement's rejection of, 23, 27–29; reactionary, 161–62; relation to leftist approaches, 19, 50–51, 106, 132, 135, 157, 162, 163, 165–67, 177; religion-science debate, 10, 15–16, 19, 167, 176–77; science wars, 19, 162–69, 174–75, 176–77, 179; scientific authority defense, 5, 8, 10, 15, 24, 27, 51–52, 68, 78, 101–2, 109, 130, 131–32, 135, 165–66; scientific neutrality, 23; sociobiology debate and, 49, 50–51, 75, 84, 89, 154, 176; Western science defense, 101, 180, 190. *See also* New Atheism

Liberty University, 113

Lloyd, Lisa, 150–51, 198n063; *The Case of the Female Orgasm,* 151

Lombroso, Cesare, 66

Longino, Helen, 71–72; *Science as Social Knowledge . . .*, 72

Lorde, Audre, 74–75, 77–78; *Sister Outside,* 75

Lorenz, Konrad, 50; *On Aggression,* 25

Lowe, Marian, 2, 5, 57–58, 70–71, 80

Luhrmann, T.M., 110

Lyell, Charles, 95

MacArthur, Robert, 155

macroevolutionary theory, 137–38

March for Science, 8

Marxism, 8, 10, 19, 30, 57

Mauldin, John, 125

Mayr, Ernst, 118, 141–42

MBL model, of history, 92–94

McClintock, Barbara, 72

McKissick, Floyd, 35

McLean v. Arkansas trial, 18, 107–9, 114–21, 122–26, 132, 136, 139, 140, 145, 147, 150, 176, 201n62, 221n68

McLemore, Monica, 155, 157

McNamara, Robert, 33

McPherson, Myra, 54

medical research, racist and sexist practices, 6–7, 11, 165

Mencken, H. L., "Scopes: Infidel," 120–21

militarism, 33, 47, 129, 130, 160, 162, 173, 177, 180, 188, 191; New Atheism and, 145, 177; post-September 11, 2001, 145, 177–78, 180, 190

military-industrial-academic complex, 8, 33, 129, 162, 168, 195n30
Miller, Peter D., 119
Mismeasure of Man, The (Gould), 2, 4, 3, 5, 12, 36, 47, 63–68, 105, 143–44, 76, 156–57, 161; feminist criticism, 69; race-IQ debate, 65–66, 74
"Mismeasure of Science . . ."(Lewis et al), 2, 4
Moallem, Minoo, 178
molecular biology, 83, 142
Montagu, Ashley, 31; *Science and Creationism*, 123–24
Moore, John, 116
morality: Christian, 112; Gould's perspective, 30, 170–71, 174; humanist, 114, 159–60; of scientists, 41; sexual, 111; sociobiological theory of, 85, 112, 170–71
Moral Majority, 113, 115–16, 121
Morgan, Thomas Hunt, 87
Morris, Desmond, 50; The *Naked Ape*, 25, 27
Morris, Henry M., 10, 127, 189; *The Genesis Flood . . .*, 113–14; *Scientific Creationism*, 114–15, 136, 196n36
Morris, Simon Conway, 215n62
Morton, Samuel George, *Crania Americana*, 1–2, 3–5, 67–68, 75, 193n1, 209n86
Muslims, 160, 175, 177–78, 180, 190

National Academy of Sciences, 25–26, 27, 117–18, 157
National Association for the Advancement of Colored People (NAACP), 30–31
National Center for Science Education, 124–26, 147, 228n80
Native American Graves Protection and Repatriation Act, 1–2
Native Americans, 1–2, 6–7, 8, 24, 78, 183, 187, 191
natural history, 55, 70
Natural History, Gould's column, 5, 13–14, 24, 30, 33–34, 37, 40, 47, 50, 56, 58–59, 64, 122, 152, 171–72, 186; Alcock's criticism of, 152–53; anthologies, 65, 151, 179; "A Visit to Dayton," 107, 119, 126; "Biological Potential *vs.* Biological Determinism," 50; "Evolution as Fact and Theory," 122–23; "Evolution's Erratic Pace," 136–37; "Freudian Slip," 151; *I Have Landed*, 179, 180; moral persuasion *vs.* collective action approaches, 41–45; popular audience, 42–43; "Racist Argument and IQ," 22, 35–36; "The Child as Man's Real Father," 42–43; "The Nonscience of Human Nature," 50, 58; "The Race Problem," 36; "This View of Life," 42, 45, 203n.92; writing style, 42–43
naturalism, scientific, 133–34, 135, 147–48, 153, 176
Naturalist, The, 158
natural sciences, 5, 18, 68, 74, 164, 165–66, 167, 170, 171; women in, 57, 59, 135
natural selection: adaptationist view, 13, 19, 81–84, 90–92, 152–53; Dawkin's perspective, 145–47; evolutionary psychology perspective, 19; Gould's perspective, 82–83, 134, 135, 136–37, 142–50; Gould's view, 82–83, 134, 135, 136–37, 142–50; impact of unpredictable events on, 83; optimality models, 87–88, 87–89, 90, 92, 94; sociobiological perspective, 19, 50, 61, 65, 83, 87–88, 94, 103, 105; taxonomic-level, 83. *See also* adaptationism; punctuated equilibria model
natural theology, 104
Nelkin, Dorothy, *Science Textbook Controversies . . .*, 119
neutrality: political and economic, 39–40, 41; scientific, 27, 36–37, 40, 48, 73, 89
New Atheism movement, 19, 86, 145, 159–60, 162, 175–80, 181
Newell, Norman, 33, 34
New Left, 8, 33, 42, 47; academic inequity practices, 163; academic institutionalization, 24, 163; Gould and, 2, 16, 17–18, 22–25, 32, 34–35, 40, 42, 174, 179, 201n62; health care activism, 23–24, 47, 79; scientific objectivity criticism, 2; student activism, 163, 174. *See also* Science for the People
New Right, 115, 122–24, 132. *See also* Christian Right
New York Academy of Sciences, 56
New York Institute for the Humanities, 74–75
Nixon, Richard, 41, 130–31
NOMA (non-overlapping magisteria), 19, 161–62, 169–75, 179
Nott, Josiah, 1
NOVA, 123
nuclear and chemical weapons, 9, 22, 32, 37,

41, 129, 130–31, 172–73, 188; antinuclear movement, 172–73, 195n32, 195n33

Obama, Barack, 45
objectivity, scientific, 2, 5, 28, 78, 80, 170; biological determinism's use of, 76; of E.O. Wilson, 54; feminist perspective, 72, 73, 75, 197n43; Gould's perspective, 66, 75, 80; humanistic perspective, 5, 182; leftist perspective, 2, 54, 75, 80, 182; liberal perspective, 3, 75, 78; racism and, 28, 40, 48. *See also* neutrality, scientific
Of Pandas and People, 147
optimality models, of natural selection, 87–89, 90, 92, 94
orgasm, female, 14, 151–54, 198n64
Our Bodies, Ourselves, 59–60, 128
Our Bodies, Ourselves collective, 59–60
Overton, William, 108, 119–20, 124
Oxford University, 146–47, 228n79; New College, 86

paleontology and paleobiology, 66, 83, 91, 145, 148, 201n62; MBL mode, 92–94; nomothetic approach, 92–94, 197n58
Palmer, Craig T., *A Natural History of Rape*, 99
parental investment theory, 85–86, 88, 98
patriarchy, 5, 9, 57, 59, 74, 79, 108, 160, 187, 191, 206n35
Paul, Diane, 4
Philippines, American imperialism in, 185–88
Pinker, Steven, 7–8, 11; *The Blank Slate . . .*, 100–101
Pitman-Hughes, Dorothy, 69
Planned Parenthood, 6–7, 77
PLOS Biology, 2, 3–4, 5
popes: John Paul II, 172; Pius XII, 172
populism: of the Christian Right, 79; of creation science, 126–32, 222n88; of leftist health movements, 79, 191; radical Black, 32–33, 47; right-wing, 79–80; of science, 3, 126–32; scientific expertise vs., 126–32
postcolonialism, 157–58, 160, 163, 165, 170
postmodernism, 10–11, 166, 167–68, 169, 176, 182
Protestantism, 15, 97, 102–3, 106, 130, 148, 172
psychology, evolutionary, 7, 12, 19, 82, 97–103; feminist criticisms, 154, 229n113; radical science's criticisms, 5
Puar, Jasbir, *Terrorist Assemblages . . .*, 177–78
punctuated equilibria model, 18, 134–35, 142, 214n49; creationism and, 18, 96, 136–40; Dawkin's criticism, 134–35, 145–47
Purcell, Rosamund, 43

queer theory/studies, 10, 73, 74, 163, 170

race: biological universalism theory, 36–37, 39–40; evolutionary psychology's perspective, 101; Gould's view of, 35, 36, 83; polygenist theory, 1, 67–68, 209n86; UNESCO statement on, 27
racial bias, influence on science, 4–5, 49, 52, 59, 68, 142, 155, 156–57; Du Bois on, 103–4; Gould's perspective, 4, 36, 40, 47, 59, 67–68, 105, 142–43, 157–58; influence on E.O. Wilson, 155–56; radical science movement's perspective, 23, 24–25; scientific neutrality and, 48
racial justice, 24, 25, 27, 29, 34, 35, 36, 40–41
racial segregation, 22–23, 27, 29, 76, 113, 114; in health care, 8, 23–24, 37, 38
racism: conservative, 39; as environmental factor, 26; intersection with technoscience, 48–49; postracialism and, 45–46; structural/systemic, 6–7, 41, 45–48, 128
racism, scientific, 5, 6–7, 23, 45–49, 132, 150, 164–65; biological determinism-based, 61; critical humanism's criticism, 163; Gould's criticism, 2, 5, 7, 17, 46, 47, 49, 52, 65–68, 97, 156–57; leftist criticism, 2, 5, 7, 15, 40, 41, 157, 160; non-Western perspective, 48–49; origin, 1–2; radical science movement's criticism, 23–25; r/K selection theory and, 155–56; scientific community's self-regulation of, 24–25; scientific liberalism's criticism, 24; as structural racism, 6–7. *See also* intelligence (IQ)-race debate
Radcliffe College, 58
Radical Science Journal, 195n32
radical science movement, 2, 8–9, 23–45, 52, 78, 160, 167; direct-action tactics, 41, 181–82, 220n48; relation to Black radical politics, 38–39; scholarly approaches, 27–30. *See also* Sociobiology Study Group

Raup, David, 92
Reagan, Ronald, 10, 41, 108–9, 111, 115–16, 123, 126–27, 172
realism, scientific, 29, 76
relativism, 15, 19–20, 29, 68–69, 131, 176, 198n64
religion-science relationship, 159, 166; Catholicism's perspective, 171–72; Gould's perspective, 19, 37, 159, 161–62, 169, 171, 174, 179; impact of September 11, 2001 on, 145, 159–60, 175, 176–79, 180, 190; leftist criticisms of science and, 160–61; New Atheism movement, 14, 19, 86, 159–60, 175–80, 181; NOMA (nonoverlapping magisteria), 19, 159, 161–62, 169–75, 179; scientific liberalism's perspective, 167, 176–77. *See also* creationism-evolution debate
"replaying the tape of life" metaphor, 96, 104–5
reproductive justice movement, 78–79
Republican Party, 41, 46, 108–9, 110–11, 112, 154, 196n42
Ridley, Matt, *The Red Queen . . .*, 100
right-wing criticisms, of science, 2–3, 131. *See also* Christian Right; conservatives/conservatism; New Right
Robbins, Bruce, 168
Roosevelt, Theodore, 77, 188
Rosenberg, Eleanor, 30
Rosenberg, Joseph Arthur, 178–79
Rosoff, Betty, 56
Ross, Andrew, 168
Roterodamus, Erasmus, 171
Ruse, Michael, 120, 146
Rushton, J. Philippe, 155–56
Russett, Cynthia, *Sexual Science . . .*, 69, 70
Rutgers University, 177–78

Sagan, Carl, 44, 123, 173
saltationism, 146
Salzman, Freda, 53
San Diego State University, 72–73
Sanford, Wendy, 59
Sanger, Margaret, 77
Sarah Lawrence College, 69–70
Scanzoni, Letha and Nancy Hardesty, *All We're Meant to Be . . .*, 102
Schaeffer, Edith, 111
Schaeffer, Francis, 111–13; *Whatever Happened to the Human Race?* 111–12
Schaeffer, Frank, 111–12
Schiebinger, Londa, *Nature's Body . . .*, 69, 70
Schlafly, Phyllis, 102
Schopf, Thomas, 92–93, 201n62; *Models in Paleobiology*, 92–93, 136
Schultz, Adolph H., 31
Schwerner, Michael, 34–35
science: Black criticism, 37–41; denaturalizing of, 11, 20; feminist criticism, 70–71, 160; leftist criticism, 8, 160–61; misuse, 9, 23, 24, 40, 61, 75; professionalization, 1–2, 4, 164; radical science movement's criticism, 23–25; scientific liberalism's perspective, 163. *See also* authority, scientific
Science, 58–59, 115, 117–18, 124–25, 137, 169
Science for the People, 23, 28, 53, 59, 78, 202n77, 205n11; Black Panther Party and, 24, 38–39; disruptive tactics, 27, 38–39, 53, 59; Gould's relationship with, 35, 42, 53, 63, 201n62, 208n72; *Science for the People* magazine, 42, 195n32
Science for Vietnam, 28
science wars, 19, 76, 162–69, 174–75, 176–77, 179
Scientific American, 127, 155, 157
scientific knowledge, 164–65, 197n50; biopolitical use, 76–78, 128–29; of creation science, 127–28; feminist perspective, 59, 69, 71–72, 73; Gould's perspective, 89, 122; radical science movement's perspective, 23, 78; religious approaches to, 106; scientist identity theory, 23, 164–65; social biases of, 5
scientific paradigms, 67; adaptationist, Gould's criticism of, 89–97; Kuhnian, 67, 89, 90, 143–44, 209n85; Panglossian, 13, 61, 81–84, 90–91, 92, 97, 129, 134, 183, 191
scientists: Black American, 103–4; identity, 23; individual beliefs, 24–25; political activism, 41. *See also* women, in science
Scopes (Monkey) trial *(The State of Tennessee v. John Thomas Scopes)*, 107–10, 118–19, 120–21, 122, 127, 218n6
Scott, Eugenie, 125–26, 147, 228n80
secularism, 3, 10, 18–19, 109, 112, 159–60, 183, 231n6

Segerstråle, Ullrica, *Defenders of the Truth,* 91–92
Sepkoski, Jack, 92
September 11, 2001, 145, 159, 175, 176–79
sex differences, theories and approaches to, 150; adaptationist *versus* development theories, 151–54; Christian evangelical perspective, 102–3; evolutionary psychology perspective, 82, 97–103; feminist perspective, 56, 58, 82, 135; Gould's perspective, 97, 151–54; leftist, women-led approach, 56–59; sociobiological perspective, 53–54, 56, 84–89, 97–98, 102, 150–51, 176. *See also* gender roles
sexism, in science, 23, 24, 58, 69, 72, 164–65; biological determinism-based, 61; critical humanism's perspective, 163; feminist criticism, 56, 58; humanistic criticism, 56; leftist criticism, 7, 157; radical science's criticism, 5, 24
sexual selection, 85–86, 885, 152, 165
Shadow Mountain megachurch, 114, 189–90
Shockley, William, 25–26, 27, 76
Signs: A Journal of Women and Culture, 70–71
Simpson, George Gaylord, 141; *The Tempo and Mode of Evolution,* 34
Skloot, Rebecca, 6
slavery and enslaved people, 1, 4–5, 6, 8, 64, 132, 164, 187
Smith, Barbara, 37–38, 39, 188
Smith, Beverly, 37–38, 39
Smithsonian Institution, 69–70
Snow, C. P., "The Two Cultures," 162
Sober, Elliot, 139
social behavior evolution, 9. *See also* sociobiology
social biases, science and, 5–6, 48, 59, 61, 66, 67, 105, 144, 163; Gould's perspective on, 12, 14–15, 65–67
social hierarchies, 5–7, 76, 78, 87, 143, 156
socialism, 38, 39
social justice, 3, 46, 59, 163, 168; feminist approach, 59, 75–76; intersectional approaches, 38; leftist approach, 8, 20, 40, 54, 163, 181–82; postmodernist approach, 10; radical science movement's approach, 27–28, 75–76; sociobiological theory of, 85
Social Text, 167–68
Society for the Study of Evolution, 116–18, 124; Education Committee, 116–17, 118
Society of Systemic Biologists, 43–44
sociobiology debate: adaptationism issue, 13, 81–84, 89–97, 104, 134, 145, 151–53, 152–53; Christian Right and, 112; evolutionary synthesis and, 142; feminist perspective, 5, 7, 12, 18, 52, 56–59, 70–71, 75, 84, 85, 99, 150, 154, 229n113; gene-centered evolutionary theory and, 86–87; Gould's perspective, 18, 81–84, 157, 170, 182; leftist perspective, 50–52, 150, 160, 176; popular media coverage, 84–85, 88; punctuated equilibria model and, 94, 96; scientific liberalism and, 54–55, 60–63, 80, 176–77; scientific racism and, 155–56; sexism issue, 51, 97; SSG and, 5, 52–55, 75, 155. *See also* adaptationism, Gould's criticism; biological determinism
Sociobiology Study Group, 5, 52–55, 75, 166
Sokal, Alan D. (Sokal affair), 167–68
Spark Reproductive Justice Now, 79
Speak Out Against the New Right, 122–23
SSE. *See* Society for the Study of Evolution
Stebbins, Ledyard, 141, 215n64
Steinem, Gloria, 69, 123
Sterelny, Kim, 139
stereotypes, sexual, 53, 55, 59, 69, 85, 89
Students for a Democratic Society, 22, 31–32, 33
Students Peace Union, 31–32
Symons, Donald, *The Evolution of Human Sexuality,* 98
Systemic Zoology, 43–44

Taylor, Keeanga-Yamahtta, "The Pivot to Class," 46–47
Ternes, Alan, 34, 41–42
Thomas, Clarence, 77
Thornhill, Randy, *A Natural History of Rape,* 99
tiers of time theory, of history, 94–96
Tobach, Ethel, 51
Trivers, Robert, 85–86, 98, 185–86
Trump, Donald, 7–8, 11, 19, 46, 126
Tuana, Nancy, 71

United National Educational Scientific and Cultural Organization, 27
US Commission on Civil Rights, 26

US Supreme Court: *Brown v. Board of Education*, 26; *Dobbs v. Jackson Women's Health Organization*, 77; *Edwards v. Aguillard*, 108; *Roe v. Wade*, 77, 128
USA PATRIOT Act, 178
University of California: Berkeley, 147, 178; Davis, 116; Irvine, 220n42; LA, BioCritical Studies Lab, 48; Riverside, 116
University of Kentucky, 117
University of Pennsylvania Museum of Archaeology and Anthropology, Morton Cranial Collection, 1–2, 3–4

Vetter, Herbert, 122–23
Vietnam War, 2, 8, 23, 30, 31, 33, 58, 71, 83, 130
Voltaire, 104; *Candide*, 82
Vonnegut, Kurt, 44

Waddington, C. H., 53, 205n11
Walters, Barbara, 114
war on terror, 160, 175, 177, 180, 189
Weinberg, Stanley, 124–25
Weisberg, Michael, 4
West, Cornel, *Prophesy Deliverance!* 40
Western science, 48–49, 101, 166; evolutionary psychology's perspective, 101; gender bias, 58; Gould's perspective, 179–80; humanistic perspective, 5; leftist perspective, 182–83; New Atheism's perspective, 175–76; racism, 25, 40; scientific liberals' perspective, 101, 180, 190; white supremacy's role, 40
Whitcomb, John C., Jr.., *The Genesis Flood . . .*, 113–14
white supremacy, 39, 40, 41, 46, 74–75, 77–78
Wilberforce, Samuel, 147, 228n79
Wilberforce University, 22, 34
Wilson, Edward O., *Sociobiology: A New Synthesis*, 9, 50–52, 80, 85, 150, 170; Christian Right's criticism, 112; feminist criticism, 56, 57; Gould's criticism, 61–65, 80; *Mismeasure of Man* as response to, 63–68; SSG's criticism, 52–55
Wilson, E.O., 61–63, 92, 109, 158, 160; *Caste and Ecology in the Social Insects*, 88; *Consilience . . .*, 170–71; evolutionary synthesis position, 142; historical criticisms of, 154–56; *On Human Nature*, 86, 99–100; scientific liberalism of, 9, 54–55, 60–63, 80, 165–66; scientific racism and, 155–56
women: of color, 6, 74–75; health movement, 37–38, 39, 49, 55–56, 59–60, 89, 99, 128; political power, 24; as religious leaders, 97, 102; suffrage, 164; Third World, 38, 74, 78, 188
women, in science, 52, 60, 71; criticism of sociobiology, 56–59, 70–71; STEM fields participation, 59, 72, 100–101, 182, 184
Women's Action alliance, 69–70
women's history, 69–70
women's movement, 9, 16, 23–24, 50, 51, 58, 69–70, 71, 76, 77, 78, 113, 187
women's studies, 58, 72–73, 163, 178
World War II, 8, 30, 87, 117, 119, 129, 162, 188
Wright, Robert, "The Accidental Creationist," 134–35, 154
W.W. Norton (publishing company), 43, 64, 65, 203n.92

Young Lords Party, 23–24, 47, 79, 128, 131, 182